TAWNY FROGMOUTH

SECOND EDITION

GISELA KAPLAN

CSIRO
PUBLISHING

A catalogue record for this book is available from the National Library of Australia

ISBN: 9781486308163 (pbk)
ISBN: 9781486308170 (epdf)
ISBN: 9781486308187 (epub)

How to cite:
Kaplan G (2018) *Tawny Frogmouth*. 2nd edn. CSIRO Publishing, Melbourne.

Published by

CSIRO Publishing
36 Gardiner Road, Clayton VIC 3168
Private Bag 10, Clayton South VIC 3169
Australia

Telephone: +61 3 9545 8400
Email: publishing.sales@csiro.au
Website: www.publish.csiro.au
Sign up to our email alerts: publish.csiro.au/earlyalert

Front cover: Tawny Frogmouth (source: Jacqueline Barkla)
Back cover: Author photo (courtesy of Prof. L. Rogers)

Set in 11/13.5 Adobe Minion Pro & Helvetica Neue LT Std
Edited by Karen Pearce
Cover design by James Kelly
Typeset by Desktop Concepts Pty Ltd, Melbourne
Printed by Ingram Lightning Source

CSIRO Publishing publishes and distributes scientific, technical and health science books, magazines and journals from Australia to a worldwide audience and conducts these activities autonomously from the research activities of the Commonwealth Scientific and Industrial Research Organisation (CSIRO). The views expressed in this publication are those of the author(s) and do not necessarily represent those of, and should not be attributed to, the publisher or CSIRO. The copyright owner shall not be liable for technical or other errors or omissions contained herein. The reader/user accepts all risks and responsibility for losses, damages, costs and other consequences resulting directly or indirectly from using this information.

CSIRO acknowledges the Traditional Owners of the lands that we live and work on across Australia and pays its respect to Elders past and present. CSIRO recognises that Aboriginal and Torres Strait Islander peoples have made and will continue to make extraordinary contributions to all aspects of Australian life including culture, economy and science. CSIRO is committed to reconciliation and demonstrating respect for Indigenous knowledge and science. The use of Western science in this publication should not be interpreted as diminishing the knowledge of plants, animals and environment from Indigenous ecological knowledge systems.

Nov25_RP_ILS

Contents

Preamble

General fascination with the tawny frogmouth has made it possible to present a revised edition of this book, first published in 2007. Since then our knowledge has expanded and tawny frogmouths, more than ever, have become a firm presence both in the psyche of Australians as well as in reality. The tawny frogmouth is the only solely Australian frogmouth and one of the two largest in the world. Its wide-ranging distribution across Australia, and the ease with which it adapts to human presence, make the tawny frogmouth perhaps the only nocturnal bird species to grace backyards from Queensland to Western Australia, from Darwin to Melbourne and beyond and, like the magpie, this species has acquired iconic status.

Despite being a largely nocturnal bird, tawny frogmouths are a much-loved species partly because they readily adapt to and accept human proximity. Any householders who have a nesting pair in their garden have found the experience delightful, particularly when a pair produces offspring that fledge in their backyards. Tawny frogmouths often seem to be indifferent to the world around them and there are some extreme examples of this. Some years ago, three rescued tawny frogmouth fledglings graced a veterinary surgery in Tamworth. Their choice of perch was an elevated part of the desk of the receptionist. Phones were ringing and animals were walked in and out of the surgery all day; I even saw a Great Dane sniffing the fledglings at a comfortable height for the dog and the birds did nothing in response. In order to speak to the receptionist and settle accounts, people had to stand nearly shoulder to shoulder with the juvenile tawny frogmouths but the birds seemed entirely unfazed by people's proximity and the commotions associated with a busy animal hospital practice. In fact, people coming into the surgery even mistook them for toys. While this was definitely a talking point within the surgery and a source of utter astonishment and amusement, the behaviour also makes good sense in terms of survival for a species with very few effective defences against predators. It is a natural adaptation and more about this behaviour will be said later. Tawny frogmouths are in fact not statues but very much alive, with a secret life that they conduct largely in the dark. It is easy to forget that this seemingly stoic and unflappable behaviour hides extremely astute and accomplished hunters, with food preferences that are seriously useful for bush and backyard.

While a good deal of research has been conducted on tawny frogmouths in recent years, there is a good deal more to learn about their behaviour.

This book was initially based on extended data collected from 36 tawny frogmouths over a 10-year period in the New England region near Armidale, New South Wales. Another 10 years have passed since the book was first published and far more work has been done, extending the initial dataset substantially both in the number of tawny frogmouths studied and in the locations used (including eastern New South Wales coastal regions). This combined work over two decades now provides the largest sample size for any study ever conducted on the behaviour of tawny frogmouths ($N = 158$). The period of observation per bird has usually spanned six weeks to three months with some follow-up of individuals. In addition, I have been privileged to follow two pairs of adults over their adult lives – one pair for six years and the other for nine years – and, in the case of one bird, for over 20 years. Such an intimate relationship with these free-living birds has allowed me to observe them at night, occasionally feeding them (taking food mid-air), and to record their behaviour and their vocalisations from vantage points nearby. I have watched them copulate, build nests, brood and raise offspring from just a few metres away, and watched their interactions and how they spend their days. The opportunity to follow them day and night, and being more than just tolerated in their presence, has led to rare observations. Their behaviour described in this book is thus the sum total of thousands of hours of observation and interaction with them over two decades.

Studying a bird of the night offers special challenges for observation and recognition but it has also given me special pleasures of learning to appreciate nights in the Australian bush. I owe these pleasures entirely to tawny frogmouths. They have not only left a profound impression but also they have been humbling because human eyes can barely adapt to the dark and here were the tawny frogmouths flying and moving through the trees with silent elegance and grace. It is to this exceptional bird that this book is devoted.

This book brings together my own extensive datasets and the relevant literature that has been published to date on tawny frogmouths, concerned with their distribution, biology, physiology, taxonomy and other relevant factors. Even taken into account are brief mentions of the 'tawny frogmouth' in some relevant context. The book also provides a rich pictorial record of the very large archive of photos that were taken during the day and also at dusk.

Finally, this preamble is to say thank you to the publisher for making this revised and updated version appear in print so that Australian readers may continue to enjoy the latest information about this iconic and very special bird. I would also like to thank the many readers who, since the first publication of the book, have become correspondents and have freely passed on their own observations, as well as images of special moments and incidents involving tawny frogmouths. These regular debates and airing of observations have led to new questions and, at times, more focused investigations on the author's part and therefore enriched the contents of the present book.

Gisela Kaplan
August 2017

1

What and where they are

Tawny frogmouths, like magpies and only a few other Australian avian species, have the distinction that they occur across Australia and only in Australia. Most people become aware of tawny frogmouths only when the birds happen to settle in their backyard as permanent residents. The perception that they are ever present may be partly an illusion created by some specific circumstances. One is that tawny frogmouths can have a very long lifespan. In London Zoo, a captive tawny frogmouth had to be euthanased recently, at the age of 32 years. One tawny frogmouth that could not be released (he has brain damage that prevents him from recognising food) was saved in 1994 and I have cared for him under licence ever since (over 23 years). When I agreed to do so, I had no idea that this meant taking on decades of care. The bird already had adult plumage when it was saved after a car accident and no definitive data were available about the lifespan of this species at the time. It will be interesting to see how long this bird will live, providing another important point of information about the potential lifespan of this species (in captivity, at least). From all available records, it appears that birds in Australia generally have lifespans that are at least twice as long as those in the northern hemisphere and often three to four times longer.

Australia's long-lived avian species sometimes give the appearance of abundance when there are in fact no new generations coming up. This has already happened in some locations concerning the kookaburra. The birds are often failing to reproduce because of lack of suitable nest holes due to the alarming decrease of tall/old trees along the entire east coast of Australia. The adults may grace a specific environment for a long time but it takes dedicated groups of birdwatchers to discover that they are not raising offspring, meaning that when the adult population dies off, there are suddenly no birds of this species left locally. Tawny frogmouths have many problems to contend with but at least they do not depend on nest holes, as kookaburras do. However, they depend on large and horizontal branches so the availability of trees that are old enough to produce such branches is also a serious matter for tawny frogmouths.

The advantage of a long lifespan and a sedentary lifestyle, at least for human observers, is that tawny frogmouths remain in one location when they have found a suitable nest site. They then form permanent partnerships and, once established in a home range, a pair may well stay for a decade or longer, unless they are subject to some misfortune. Hence, when tawny frogmouths settle in a backyard, there is plenty of time to get to know them. They may even build a nest in the same tree and on the same branch year after year (called site fidelity), and thus become part of human consciousness. Although not quite having the same iconic status as a magpie or a kookaburra, the tawny frogmouth is loved and enjoyed by many, especially on the urban fringe and in rural areas of Australia, as will be discussed in more detail later.

Attitudes to tawny frogmouths

Birds of the night tend to be surrounded by mystery, myths and legends because darkness hides most of their active behaviour from curious human eyes and that is true of owls as well as of tawny frogmouths. Apparently, some Aboriginal nations feared the hoots of the tawny frogmouth at night as an omen of their own impending death but mostly such powers of doom (and wisdom) have been reserved for owls.

Nevertheless, the tawny frogmouth has not been treated all that kindly in the past by taxonomists and ornithologists. In a recent magazine story, it was even claimed that 'the prize for the world's most unfortunate-looking bird must surely go to that golden-eyed, lock-jawed frogmouth'.[1] Calling a bird a 'frogmouth' in the first place is hardly complimentary. Certainly, the tawny frogmouth, like all frogmouths, has a wide and strong beak and a gape (beak opening) that is unparalleled in size by any bird but bristles and feathers are draped around this beak so nicely and attractively that the connotation of 'frogmouth' seems thoroughly undeserved in my opinion.

Their Latin name *Podargus strigoides* also attaches some negative meaning to frogmouths. '*Podargus*' is the Latin term for a 'gouty old man' and that, presumably, refers to their short legs and their hobbling, shuffling walk. '*Strigoides*', meaning 'owl-like', is a little more scientific. Ingram[2] tried to rescue the honour of the tawny frogmouth by pointing out that Hector's horse in Homer's epic

The Iliad was named Podargus and meant 'swift-footed' but, unfortunately, 'argos', in Greek, means slow/lazy. Clearly, tawny frogmouths are not swift-footed. The *Handbook of Australian, New Zealand and Antarctic Birds*, Vol. 4 (referred to hereafter as *HANZAB*),[3] deals with tawny frogmouths and nightjars and has a few more interpretations and translations of the name. For instance, it cites Georges Cuvier, who named frogmouths 'podarge' with reference to their short legs and supposedly weak toes. Until the 1960s there were still those who took it upon themselves to 'judge' Australian birds as inferior (compared to European birds, presumably) and to this day, I do not understand why someone would bother to write about an Australian bird merely to denigrate it.

Hence, the tawny frogmouth was introduced to the Australian public as 'grotesque', ugly, weak-footed and altogether stupid and silly.[4] Having hand-raised them and seen them in their natural habitat, I wonder how such an assessment was objectively possible or subjectively defensible. Superbly equipped for its niche, the tawny frogmouth is one of the largest of Australia's night birds (along with the powerful and rufous owls). It is a graceful flyer and to us, and to agriculture, extremely useful. It feeds on just about everything we regard as pests. Moreover, as individuals, tawny frogmouths often have a charming disposition, can be gentle, tender (Fig. 1.1) and even emotional, similar to parrots, and are avid communicators and stubborn as mules at times. In other words, they have strong individual personalities and are very clever in what they do.

The person to whom it fell to name this Australian species was Colonel William Legge who, in a meeting of the Australasian Association for the Advancement of Science in 1895 in Brisbane, chose the name frogmouth because of the link to its Asian relatives. *HANZAB* lists no fewer than 39 Aboriginal names for the tawny frogmouth, which attests to the fact that tawny frogmouths have been widespread across the Australian continent for some considerable time and have been familiar to many Aboriginal nations speaking very different languages.

Other common names for the order to which frogmouths belong, but not usually known and used in Australia, are 'goatsuckers'. This even more peculiar name for a frogmouth stems from ancient myths purporting that such birds milked goats or sucked their blood at night! This mythology might have arisen either in Central or South America or in Asia but must have been equally alive in Europe or its Latin name would not have been chosen to be *Caprimulgus*, the Latin name for goatsucker (by Swedish biologist Linnaeus in 1758). In northern South America, particularly Venezuela, the oilbird, *Steatornis caripensis,* a relative of the tawny frogmouth, is noted for living in dark caves and finding its way home by echolocation, similar to bats. The association of bat and bird, and the strange clicking noises of their sonar that, unlike that of bats, operates partly within human hearing range, apparently led locals to associate the two very dissimilar cave dwellers. Oilbirds are actually fruit-eaters (the only such species among this

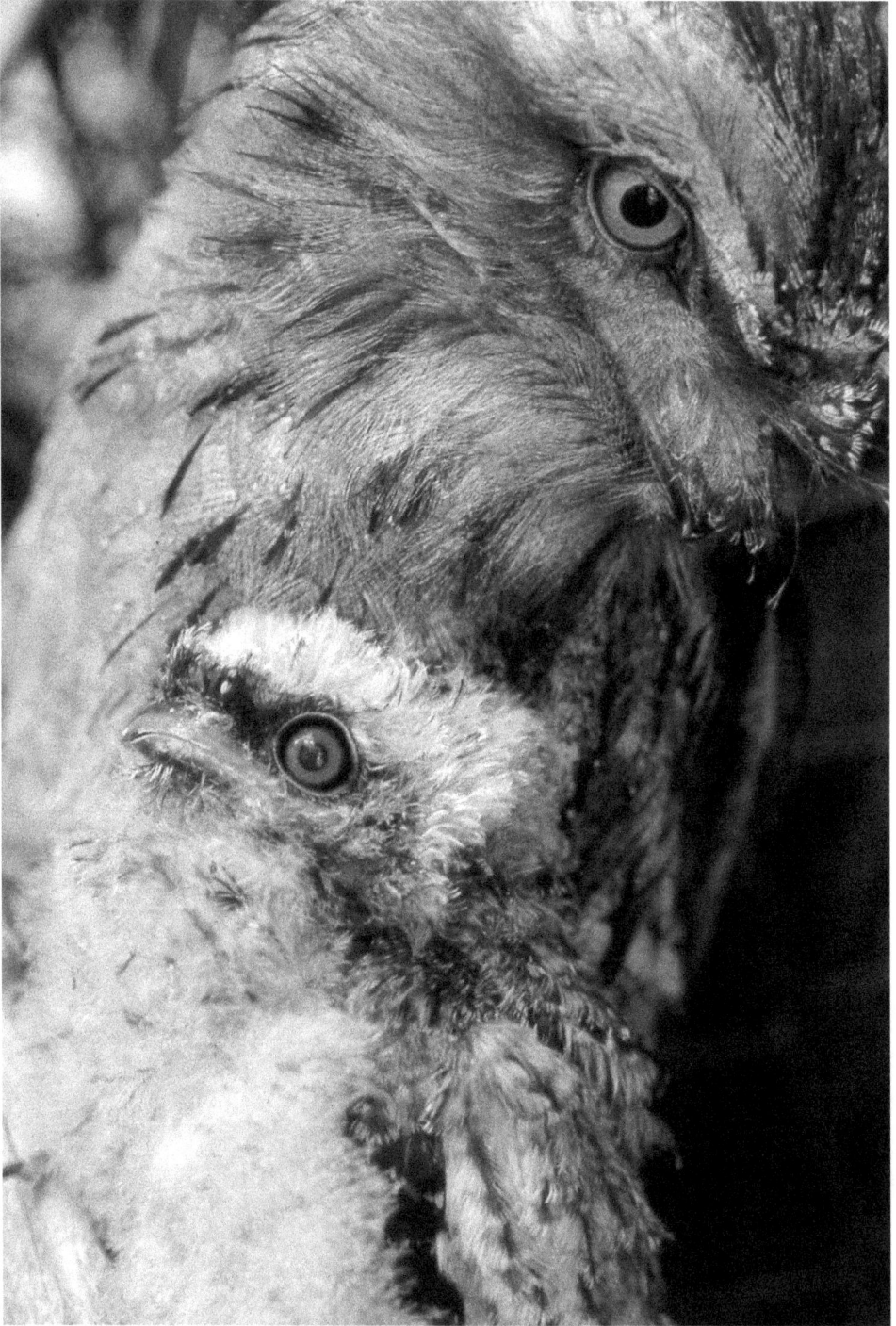

Fig. 1.1. The male is showing exceptional tenderness towards his offspring. He provides protection and equally contributes to brooding and feeding, making one realise that tawny frogmouths may have been misjudged and underrated.

order of Caprimulgiformes) and do no harm but some of the bats living in close vicinity of the oilbirds in Venezuela are vampire bats and they will certainly plague goats and other livestock, not for their milk but for their blood. Oilbirds and some of the Asian species of frogmouth also use loud and eerie calls. Apparently, the calls of some Asian species are so weird that locals have been known to have a superstitious dread of them.[5]

Who are the relatives of the tawny frogmouth?

Tawny frogmouths belong to the order Caprimulgiformes with five families, all of them nocturnal birds or crepuscular (twilight) feeders. Families in the order Caprimulgiformes (birds that are related to the frogmouth) include: oilbirds (Steatornithidae) of northern South America and Trinidad; potoos (Nyctibiidae) distributed from Argentina to Mexico; owlet-nightjars (Aegothelidae) of Australia and owlet frogmouths of New Guinea; and the relatively large family of nightjars (Caprimulgidae) found across all five continents. Which of the families is most closely related to frogmouths has been a matter of dispute.

The fifth family within the order Caprimulgiformes is that of the frogmouths proper, the Podargidae. Tawny frogmouths are one of 14 species of frogmouth belonging to two genera, *Podargus* (three Australian species) and *Batrachostomus* (11 Asian species). Ten of the species that live in Asian regions are tropical birds and none of these are found in Australia. The three *Podargus* species found within Australia are the tawny frogmouth (*Podargus strigoides*), the marbled frogmouth (*Podargus ocellatus*) and the Papuan frogmouth (*Podargus papuensis*). Two of these, the marbled frogmouth and the Papuan frogmouth, are tropical birds and their habitat extends to New Guinea. The Papuan frogmouth occurs only on Cape York Peninsula in the far north of Australia and, as the name implies, is at home in Papua New Guinea. It has red eyes and, outside Australia, is supposed to be even larger than the tawny frogmouth (up to 54 cm, nearly the size of the powerful owl, one of the largest owls in the world). The marbled frogmouth is found in two disparate populations, sharing part of its distribution with the Papuan frogmouth on Cape York Peninsula but a second population occurs in coastal and near-coastal southern Queensland and northern New South Wales (see Fig. 1.2). I have spotted one in a rainforest area in the hinterland of Woolgoolga, some 50 km further south than any surveys to date have recognised (none have been conducted there). The bird was spotted easily because it roosted on a tree stump, quite exposed to the onlooker, and did so during daytime. This may have been a singular event of sighting a stray but some of these areas have simply not been surveyed. Nevertheless, marbled frogmouths are quite rare in Australia and most of them live in Papua New Guinea. Indeed, worryingly, *The New Atlas of Australian Birds*[6] has not been able to identify a single breeding site for this species within Australia. However, recently, a new

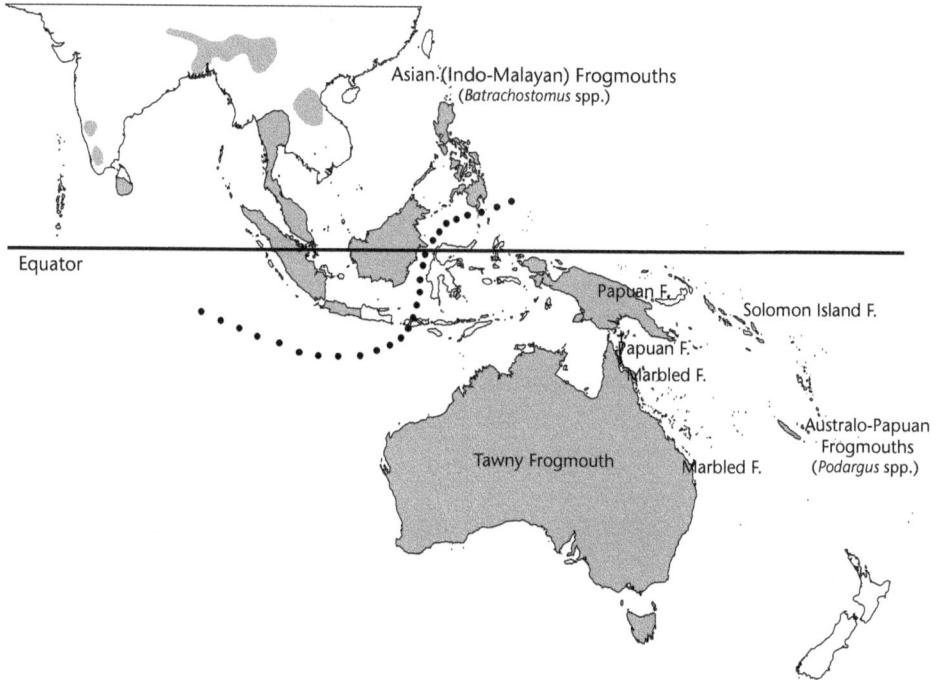

Fig. 1.2. Distribution of all frogmouth species and their allies (shaded areas). The black line indicates the equator; the dotted line is the Wallace line and shows the separation of Australian/Papuan frogmouths (*Podargus*) and Asian frogmouths (*Batrachostomus*). At its narrowest point (between the islands of Bali to the west and Lombok to the east) there are only 25 km. The distribution of the Solomon frogmouth (Solomon Islands east of Papua New Guinea) is shown by shading the specific islands. Not shown are the frogmouths of central and northern South America.

species of *Podargus* has been described, based on former identification of the Solomon Island frogmouth.[7]

And then there are the Asian relatives, the *Batrachostomus* species, living north of the so-called Wallace line (dotted line in Fig. 1.2), currently constituting a separate family, Batrachostomidae. The Wallace line is an imaginary line drawn by Alfred Russel Wallace (1823–1913), an explorer contemporary of Charles Darwin, based on his observations that neither flora nor fauna seemed to cross this line, probably due to deep water channels.[8] This line is drawn along the Makassar Strait separating the flora and fauna of Asia from the flora and fauna of the islands to the east of it. To the western side of it there are the Asiatic plants and animals of mainland Asia, Borneo, Sumatra, Java and Bali and to the eastern side of it are many different species on the islands of Lombok and other smaller islands. To date, the Wallace line seems to hold for many mammals but, of course, many avian species are competent long-distance flyers. One of the most obvious absences on the Australian and Papuan shelf are primates, and one of the most obvious absences north-west of the Wallace line are marsupials.[3] The dotted line in Fig. 1.2

indicates the separation of flora and fauna north-west and south-east of the Wallace line. In the case of frogmouths, it also indicates the separation of Batrachostomidae from Podargidae. Of late, it has been argued that frogmouths may have a Gondwanan origin and that Batrachostomidae split off from the Australo-Papuan Podargidae.[7]

The tawny frogmouth is unique to Australia. Notwithstanding the number of species related to it, other frogmouths are rare in Australia, and all other species in Asia are widely separated. The Australian tawny frogmouth is one of only two temperate frogmouth species (the other lives in the Himalayas) and the only one of 14 species that has learned to live with and survive night frosts and long stretches of very limited food supplies.

Complexities of subspecies, geographic variation and systematics

In the past, taxonomists and ornithologists have grappled with some local variations of colour, size and other differences in tawny frogmouths but they came to the conclusion that these were not geographic variations but actually subspecies of the tawny frogmouth. At one time, there were ~24 different scientific names given to tawny frogmouths. Other writers recognised between five and seven different subspecies,[9] justified by pointing out slight differences in plumage colour and size. It is a little confusing and rather pointless to tell this whole story[10] because taxonomists themselves recently acted to simplify matters and reduce the number of recognised subspecies.

Currently four subspecies are recognised. *Podargus strigoides strigoides* is basically a south-eastern and eastern Australian bird occurring from Townsville (Queensland) to Victoria, along the southern side of the continent and as far as Spencer Gulf in South Australia, although the exact limits of their distribution are still under debate. They are the largest, darkest and, in plumage colour, the most varied form. Sexual dimorphism has been claimed to be strong in *P. s. strigoides*, largely in terms of bill size, wing and tail length and also in colour morph, but this is perhaps somewhat exaggerated because, in reality, they are difficult to tell apart in the field.

According to my own data, weight should be added as an important index of sexual difference. Weights of tawny frogmouths are indicated in *HANZAB*; however, my own measurements of weight do not conform to the weights published in this volume. Weights of almost all tawny frogmouths in my dataset – and every single one was weighed on arrival and departure – are sometimes considerably higher than *HANZAB* indicates (Fig. 1.3). Weight ranges for males are listed as 178–520 g. The heaviest tawny frogmouth male in my list weighed 750 g and most were between 440 and 600 g. Female weights are listed as 157–555 g. The weights at

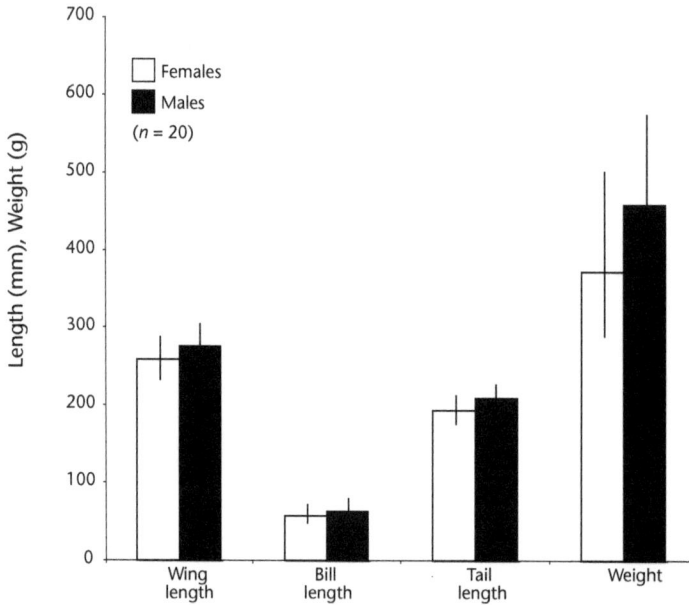

Fig. 1.3. Size dimorphism showing mean differences in wing, bill and tail sizes. These differences show some overlap. In weight (right column) the sexual differences are most pronounced.

the low end of the scale for both adult male and female *P. strigoides* in this published account surprised me. Not one of the tawny frogmouths in my encounters has ever weighed that little. The lowest weight for a female was 300 g and the lowest for a male was 350 g, and these were birds showing signs of dehydration and starvation. Whatever caused these significant weight differences (wide geographic variation may be the reason) does not change or undermine the assumption that tawny frogmouths are sexually dimorphic at least in weight or that there are large variations in size within subspecies (up to 200% greater weight and 40% longer wing length[9] for *P. s. strigoides* than for the northern type).

In the field, the sexes are difficult to distinguish unless one sees a male and a female together. Even together, it takes some careful judgment to spot the differences. Two tawny frogmouths sitting together may adopt different postures and it is almost impossible to tell at times which one is the larger. Perhaps the clearest sex difference is in the size and shape of the beak (Fig. 1.4). The male's beak seems flatter, more exposed and more triangular than the beak of the female, although there is a great deal of overlap in beak shape and size as well and, of course, one needs to be close enough to identify differences in the size of a beak. It is different, of course, when a dead tawny frogmouth is autopsied and sexed. Only in this manner has certainty on this issue arisen.

Another subspecies, *P. s. brachypterus*, is the western variety, found from west of the Great Divide to Western Australia. It only occurs in grey forms and sexual dimorphism is only weakly apparent. Then there are those tawny frogmouths

Fig. 1.4. In the field, it can be difficult to distinguish males from females because of postures adopted by the bird. Here, the male is on the left (note the wider bill) and the female on the right.

named *P. s. phalaenoides* of the tropical north of Australia with rufous and grey morphs for females and paler grey than east coast tawny frogmouths. To make matters more complicated, *P. s. strigoides*, *P. s. brachypterus* and *P. s. phalaenoides* all meet in north-eastern Queensland and interbreed to give taxonomists a headache. Finally, there is the subspecies named *P. s. lilae* but this occurs only on Groote Eylandt, off the coast of the Northern Territory in the Gulf of Carpentaria. This is the smallest of all, uniformly silvery grey, less patterned and showing little sexual dimorphism. In general, it is said that the further north (towards the tropics) one goes, the smaller tawny frogmouths become. However, there is some confusion about sizes generally. Hollands,[11] for instance, thought that male tawny frogmouths are generally smaller than females but, as described above, the opposite is the case.[3] Across all subspecies, there are also differences in eye colour, ranging from pale yellow to deep orange, but there is considerable overlap. Note that *The New Atlas of Australian Birds*[6] made no such distinctions between subspecies but called all tawny frogmouths by the one name, *Podargus strigoides*.

Taxonomically, the position of the tawny frogmouth and its affiliations are not all too clear either. A quiet revolution has taken place in taxonomy in the last decades. It is now confirmed that all songbirds and several other lineages of

Peters	Sibley & Monroe	Livezey & Zusi	
PAS	PAS	PAS	Passerines
PSI	PSI	PSI	Parrots
FAL	CIC	FAL	Falcons
GRU	GRU	GRU	Seriemas
PIC	GLB	PIC	Puffbirds
PIC	PIC	PIC	Honeyguides, woodpeckers
COR	COR	COR	Kingfishers, bee-eaters
COR	BUC	COR	Hornbills
COR	UPU	COR	Hoopoes
TRO	TRO	TRO	Trogons
COR	COR	COR	Cuckoo-rollers
STR	STR	STR	Owls
COL	COL	COL	Mousebirds
FAL	CIC	FAL	New world vultures
FAL	CIC	FAL	Hawks, eagles

Other avian groups (not shown)

APO	TRC	APO	Hummingbirds
APO	APO	APO	Swifts
CAP	STR	CAP	Owlet-nightjars
CAP	STR	CAP	Frogmouths
CAP	STR	CAP	Nightjars
CAP	STR	CAP	Oilbird
CAP	STR	CAP	Potoos

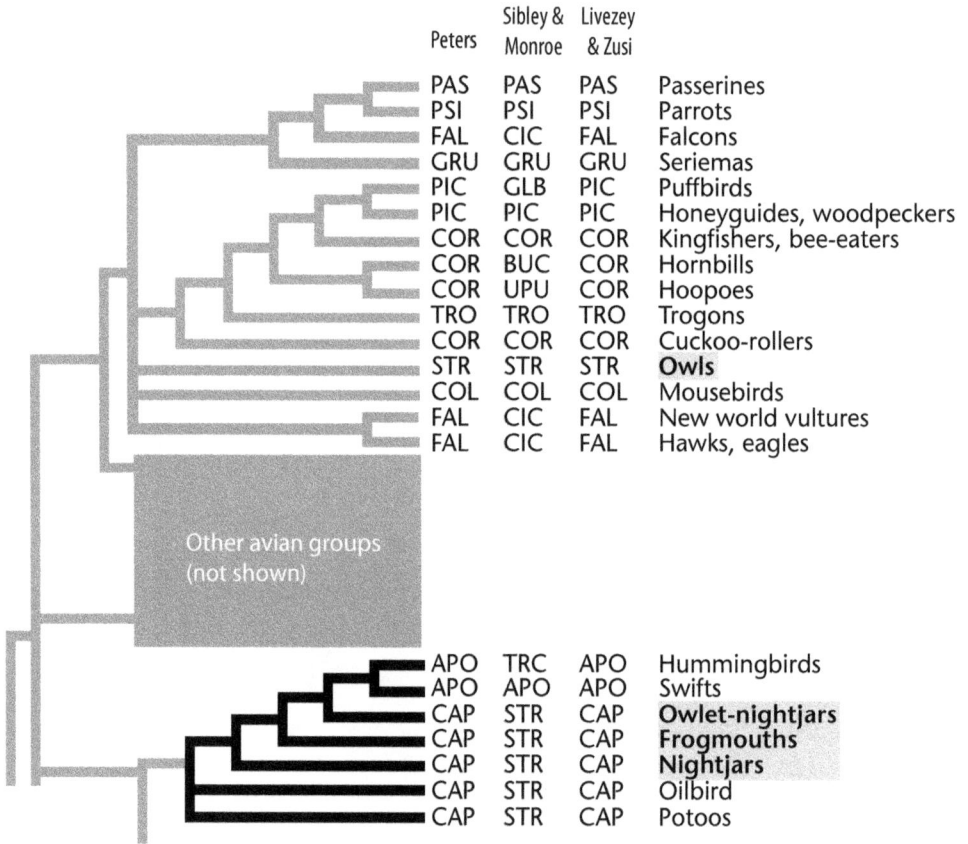

Fig. 1.5. The bars refer to relatedness and evolution of specific families. The light grey coloured tree contains the owls, while the black tree indicates the position of frogmouths and nightjars. The grey box indicates the substantial numbers of different avian groups that separate owls from frogmouths but the families themselves are not relevant here. The vertical columns indicate suggestions on where avian families belong, showing considerable disagreement between three different taxonomists. The abbreviations for orders are: APO (Apodiformes), BUC (Bucerotiformes), CAP (Caprimulgiformes), CIC (Ciconiiformes), COL (Coliiformes), COR (Coraciiformes), FAL (Falconiformes), GLB (Galbuliformes), GRU (Gruiformes), PAS (Passeriformes), PIC (Piciformes), PSI (Psittaciformes), STR (Strigiformes), TRC (Trochiliformes), TRO (Trogoniformes), and UPU (Upupiformes). The figure is a pictorial summary simplified and adapted from Hackett *et al.*[13] The three taxonomies merely serve as example here (there are many more proposals) and are far more extensively discussed by Hackett *et al.*[13]

modern birds are of Gondwanan origin (more precisely from East Gondwana, now Australia). It is now believed that several avian lineages survived the mass extinction events of 65 million years ago[12] and led to the evolution of all modern perching birds and a range of other superfamilies and clades.

Debates have been very lively about the evolution of species that do not belong to the songbirds, parrots, fowl or emus and others. The frogmouths are actually one group that now seems entirely unresolved. It does not help that fossil remains, especially of forest dwelling birds, are more or less entirely non-existent and data have had to be inferred from DNA, morphology and a variety of other methods.

Indeed, taxonomists admit that some deep avian evolutionary relationships have been difficult to resolve and, interestingly, the tawny frogmouths belong to one of those groups. Hackett *et al.*[13] gave examples of this pertaining to the Caprimulgiformes (the group to which frogmouths belong). A summary of their comparison illustrates the problem (see Fig. 1.5).

In three reconstructions of the avian tree of life, the group of birds we call nightjars are either placed with owls (Strigiformes) or within the group of birds of nightjars (called Caprimulgiformes) whose ancestry and origins are entirely separate and different from owls (Table 1.1). The question, to this day, is whether tawny frogmouths are a specific kind of owl or whether they evolved entirely separately and from different origins. In appearance, there are certainly several features that make them look similar to owls, as will be discussed later. Some of the nightjars, such as potoos, have even been called 'near passerines' – they too have three toes forward and one backwards as do passerines.

The problem is that there is an additional factor in evolution called convergent evolution, that is, the evolution of certain traits in an organism due to environmental pressures. Expressed differently, traits may evolve and be lost in quite different and unrelated species and even orders, depending on how useful or essential they are to an organism's survival. Frogmouths that are largely crepuscular (twilight) hunters, for instance, benefit from having large eyes facing forward and from stereopsis, that is, the ability to have binocular vision and hence three dimensional sharp images at some distance. Such a feature by itself does not make the frogmouth an owl, of course. Most phylogenetic studies show that frogmouths are not related, let alone closely related, to owls but not all do.

Significantly, frogmouths catch prey with their beaks, not their feet and thus lack the talons typical of birds of prey. In other words, the matter of where tawny frogmouths belong in the avian tree remains unresolved.

Distribution of tawny frogmouths within Australia

Some of the many bird surveys conducted throughout the 20th century have not only identified the presence of tawny frogmouths but also indicated whether they were nesting or not. The survey data have been well summarised in *HANZAB*,[3] and the *New Atlas of Australian Birds*[6] has also identified where tawny frogmouths have been sighted and where they have nested in broad geographical terms. They are sparse or non-existent in inland deserts, rarely reported from some of the open country and treeless deserts[15] and in Kakadu National Park they were found in open forest, edge and spinifex woodland, but were missing from the monsoon

Table 1.1 Nightjar and frogmouth family groups

Species name	Common name	Distribution	Reference
Australian and New Guinea regions			
Podargus ocellatus	Marbled frogmouth	New Guinea, NE Australia	Quoy & Gaimard 1832
Podargus papuensis	Papuan frogmouth	New Guinea, N Australia	Quoy & Gaimard 1832
Podargus strigoides	Tawny frogmouth	Australia	(Latham 1801)
Eurostopodus argus[A]	Spotted nightjar	Australia	Hartert 1892
Eurostopodus mystacalis	White-throated nightjar	Australia, winters in New Guinea	(Temminck 1826)
Eurostopodus argus	Spotted nightjar	Australia	(Hartert 1892)
Caprimulgus macrurus[B]	Large-tailed nightjar	Australia	Horsfield 1821
Aegotheles cristatus[C]	Australian owlet-nightjar	Australia, S New Guinea	(Shaw 1790)
SE Asia, Malaysian and Oceanic regions			
Batrachostomus septimus	Philippine frogmouth	Philippines	Tweeddale 1877
Batrachostomus stellatus	Gould's frogmouth	Malay Peninsula, Sumatra, Borneo	(Gould 1837)
Batrachostomus moniliger	Sri Lanka frogmouth	SW India, Sri Lanka	Blyth 1849
Batrachostomus hodgsoni	Hodgson's frogmouth	NE India to SE Asia	(Gray 1859)
Batrachostomus poliolophus	Short-tailed frogmouth	Sumatra	Hartert 1892
Batrachostomus mixtus	Bornean frogmouth	Borneo	Sharpe 1892
Batrachostomus javensis	Javan frogmouth	Java, SE Asia, Borneo	(Horsfield 1821)
Batrachostomus affinis	Blyth's frogmouth	Banggi Island (N of Borneo)	Blyth 1847
Batrachostomus chaseni	Palawan frogmouth	Palawan (Philippines)	Stresemann 1937
Batrachostomus cornutus	Sunda frogmouth	Sumatra, Borneo	(Temminck 1822)
Eurostopodus nigripennis	Solomons nightjar	Solomon Islands	Ramsay 1882
Eurostopodus exul	New Caledonian nightjar	New Caledonia	Mayr 1941
Eurostopodus diabolicus	Satanic nightjar	Sulawesi	Stresemann 1931
Eurostopodus papuensis	Papuan nightjar	New Guinea	(Schlegel 1866)
Eurostopodus archboldi	Archbold's nightjar	New Guinea	(Mayr & Rand 1935)
Lyncornis temminckii	Malaysian eared nightjar	Malay Peninsula, Sumatra, Borneo	Gould 1838

Species name	Common name	Distribution	Reference
Lyncornis macrotis	Great eared nightjar	widespread, also Sulawesi	(Vigors 1831)
Gactornis enarratus	Collared nightjar	Madagascar	(Gray 1871)
Batrachostomus auritus	Large frogmouth	Malay Peninsula, Sumatra, Borneo	(Gray 1829)
Rigidipenna inexpectata	Solomons frogmouth	Solomon Islands	(Hartert 1901)
Eurostopodus nigripennis	Solomons nightjar	Solomon Islands	Ramsay 1882
Batrachostomus harterti	Dulit frogmouth	Borneo	Sharpe 1892
Central and South America regions			
Steatornis caripensis	Oilbird	N, NW Central America, also Panama	von Humboldt 1817
Nyctibius grandis	Great potoo	S Mexico through to Amazonia, and SE Brazil	(Gmelin 1789)
Nyctibius aethereus	Long-tailed potoo	From Amazonia to SE Paraguay	(zu Wied-Neuwied 1820)
Nyctibius jamaicensis	Northern potoo	Mexico to Costa Rica, also Greater Antilles	(Gmelin 1789)
Nyctibius griseus	Common potoo	Nicaragua to Uruguay	(Gmelin 1789)
Nyctibius maculosus	Andean potoo	Venezuela to Bolivia	Ridgway 1912
Nyctibius leucopterus	White-winged potoo	Amazonia	(zu Wied-Neuwied 1821)
Nyctibius bracteatus	Rufous potoo	Amazonia	Gould 1846

The names in brackets in the last column refer to authors who first described a species.[14]
[A] In the genus *Eurostopodus*, three nightjar species are in Australia and another five are island populations of the Coral Sea and in New Guinea.
[B] In the genus *Caprimulgus* containing some 39 species (not listed), only one is named so in Australia.
[C] In the genus *Aegotheles* containing nine species (not listed) only one (owlet nightjar) is represented in Australia.

rainforest.[16] Tawny frogmouths are found along river courses if timbered[17] and in Western Australia, they are particularly numerous in areas with good growth of river gums and casuarinas.[18]

Several reports have also confirmed that tawny frogmouths have adapted to human presence and can be found nesting in garden trees and parks. They are common in suburbs[19] and have been seen feeding on moths attracted to street lights.[17] Recent surveys confirm the presence of tawny frogmouths in towns and the lure of street lights as an assembly point for flying insects (Fig. 1.6) is a ready food source.[20] This close proximity to human habitation is not just a recent development in the roosting and feeding habits of tawny frogmouths.

However, a very detailed study about the presence of boobook owls, owlet-nightjars and tawny frogmouths across a suburban-forest gradient found that the number of tawny frogmouths actually increased in response to increasing levels of

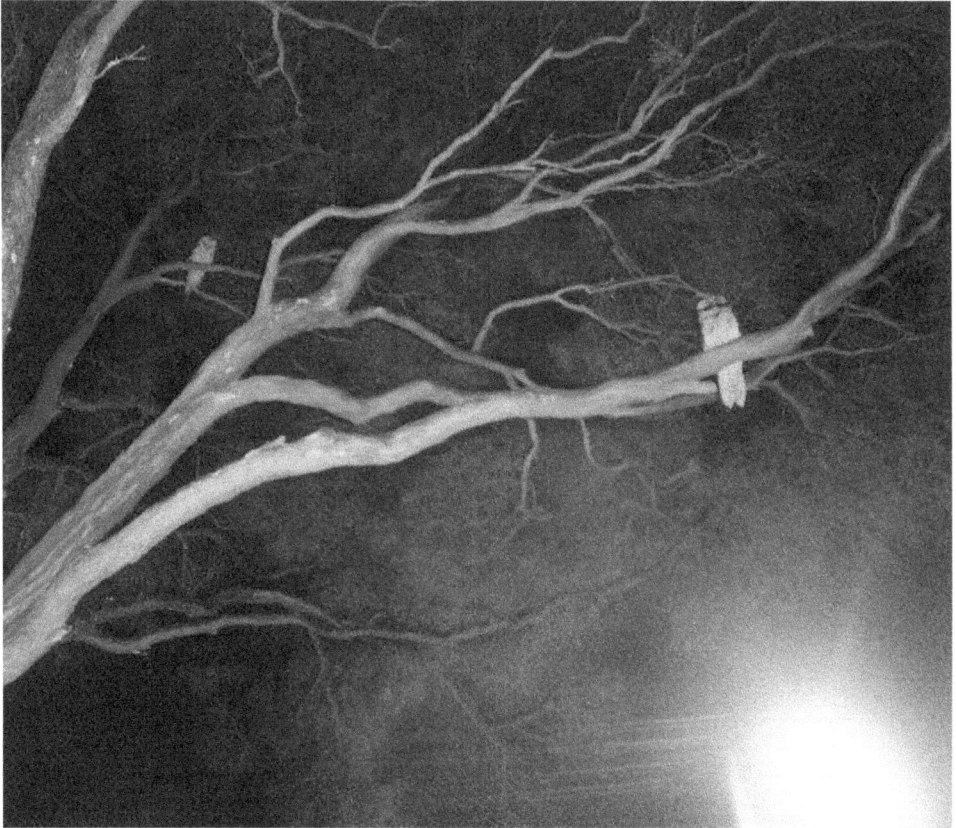

Fig. 1.6. Tawny frogmouths are attracted to street lights when moths assemble around such lighting. Here are two tawny frogmouths, one far on the left, one near on the right side of the tree; the street light is just below the tree.

urbanisation while that of the owlet-nightjars and southern boobook owls decreased. The authors explained this in terms of difference with more generalist habitat requirements. Tawny frogmouths, as said before, need good trees but both the nightjar and the owl require nest holes, which are much more difficult to find and presuppose the presence of trees that are usually more than 100 years old.[21] A later study of tawny frogmouth habitat confirmed the increased presence in urban areas, with a few important qualifications: as long as the surfaces were not all impervious (meaning concreted in or covered in bitumen), but had some grass, parks and open areas as well as large enough trees for day roosts and nest building. In other words, the tawny frogmouths aligned their home-range to the least urbanised areas of the broader urban landscape. Interestingly, it seems that in non-breeding times females tend to stay closer to the fringe while males tend to expand more into the urban landscape.[22] The same researchers also measured success of settlement in the urban fringe by breeding success and that was extremely high. In 133 breeding attempts, 177 chicks were successfully fledged.[23] However, this success has to be tempered by the fact that tawny frogmouth nestlings and first week fledglings regularly and in high numbers turn up orphaned or stranded, suggesting that a high number may be lost to various misadventures (more of this later).

Relative altitude of the terrain is also often a dividing factor in habitat selection.[24] It is interesting to look at these maps in detail. Three transparencies were provided and once superimposed upon each other a clear pattern emerges, which indicates that in the area surveyed (all environmental/habitat conditions being equal) tawny frogmouths had distinctly higher densities at altitude – not at sea level but at least at 100 m and up to 250 m above sea level. Another study at Bega in New South Wales found that they were present in gullies and ridges but more often on ridges.[25] However, these reports do not tell us whether extensive land clearing in the lowlands had occurred and had forced the tawny frogmouth to higher ground, in which case their abundance at higher altitude may not be an expression of habitat choice. Tawny frogmouths do not appear to like the really high altitudes that have snow cover in winter, judging by a survey conducted of the Thredbo Valley of Kosciuszko National Park.[26] The report concluded that tawny frogmouths were rare and their calls not heard at night. It is also possible, of course, that certain habitat characteristics discourage or foster an abundance of predators and altitude is only secondary or incidental to predator presence.

If one pieces together the information from surveys[27] it appears that, in summary, tawny frogmouths prefer relatively open wooded habitat and are found at edges of forests and of cleared areas.[28] Higher tawny frogmouth densities are found in regions of slight altitude with some good tree cover. Areas that are converted for agricultural use and are clear-felled displace tawny frogmouths entirely as has happened, for instance, from the various wheat belts of western and eastern Australia[29] but in many of those areas they have moved closer to towns.

Overall, tawny frogmouths show remarkable adaptability to many climate zones and habitats and it seems that their anatomy (Chapter 2) and physiology (Chapter 4) are part of the explanation. Moreover, as generalists in the insectivorous range of bird foods, all Australian habitats for most of the time can provide enough to sustain even a large bird like the tawny frogmouth (see Chapter 5).

2

General anatomy

Birds, if this needs to be said at all, share certain anatomical features such as two wings, feathers, two legs and a beak; these traits are not found together in any other class of animals. There are, of course, differences in detail and their specific features are adaptations to the way they feed and to the habitat they occupy. These finer points of design can be remarkable and indeed surprising, and there is much that human technology and innovation owes to birds. The tawny frogmouth has quite a few unusual and special features, so this chapter will largely confine itself to those, although some general features will be covered as well.

Integument

The integument is the name given to all external surfaces of a bird, including its skin, feathers, the bill and any protrusions. It protects the bird from external adverse objects and conditions and provides a protective shield for the sensory organs. The most conspicuous and extensive part of the integument is the feather cover.

The skin of birds in general is fragile and a good deal thinner than that of mammals. It adheres more to the bones than to other connective tissues as is the case in mammals. Muscles are interconnected by tendons and they connect to feathers. The skin of tawny frogmouths has no sweat glands, and no true cutaneous glands, except for glands in the external ear and around the vent, producing a mucus.

Most birds have a gland situated dorsally near the tip of the tail, called the preen gland or uropygial gland. The preen gland produces a fluid (lipoid sebaceous material) that a bird beak normally collects during preening and then wipes on feathers for waterproofing. In tawny frogmouths, this gland is vestigial and largely non-functional. Yet tawny frogmouths are among the best waterproofed birds, apart from waterbirds, that I have ever come across. Even after severe rain, tawny frogmouths, sitting it out in the open, at best might show a droplet hanging in the tufts of their feathers above the beak. They are otherwise completely dry, even on their heads where most birds show some signs of waterlogging!

Feathers

How do tawny frogmouths achieve such effective waterproofing without the use of a preening gland? The answer lies in a type of feather referred to as a powder-down feather. This type of powder-down feather sheds a fine waxy powder. The powder consists of tiny granules of keratin ~1 mm in diameter. We know that some parrots, bowerbirds, pigeons and herons also possess powder-down feathers.[1] They are usually similar in structure to down feathers but they can also be semi-plumes, and all contour feathers (see Fig. 2.1) produce small amounts of powder. In tawny frogmouths, it is not known whether these feathers with the highest keratin production are scattered, as in parrots, or bunched together in patches, as in herons. Femoral powder-downs seem to be the main keratin source[2] and are certainly effective.

Feathers are enormously versatile. They protect birds from heat and cold, sunlight and rain, as well as enabling them to fly (see also the section on wings in this chapter). The position of the feathers (sleeked down, fluffed or raised) helps a bird to regulate its body temperature. Feathers often have the additional function of providing camouflage (called cryptic plumage). Tawny frogmouth plumage is subdued, in greys and rufous tones, as in nocturnal birds anywhere in the world. Shades of grey may work very well as camouflage, not only for the purpose of protection from predators but also as camouflage for a predator.

Tawny frogmouths have all of the types of feathers that birds have developed. They have down feathers, filoplumes, bristles, semi-plumes and, as in most other birds, they have flight feathers (remiges), tail feathers (retrices) and contour feathers. Hatchlings and nestlings are richly endowed with white downy feathers and these are already very long and cover the whole body at the time of hatching (more in Chapter 7). From that initial down, tawny frogmouths rapidly develop juvenile plumage but some paired down remains into adulthood (in the lumbar region). Juvenile plumage on the body and even the head is particularly soft and thickly layered (Figs 2.1 and 2.2).

These thick layers of very soft plumage insulate the young birds exceptionally well from cold and heat and possibly also from ectoparasites and skin infections. Indeed, all birds that I have raised have been visually examined for ectoparasites

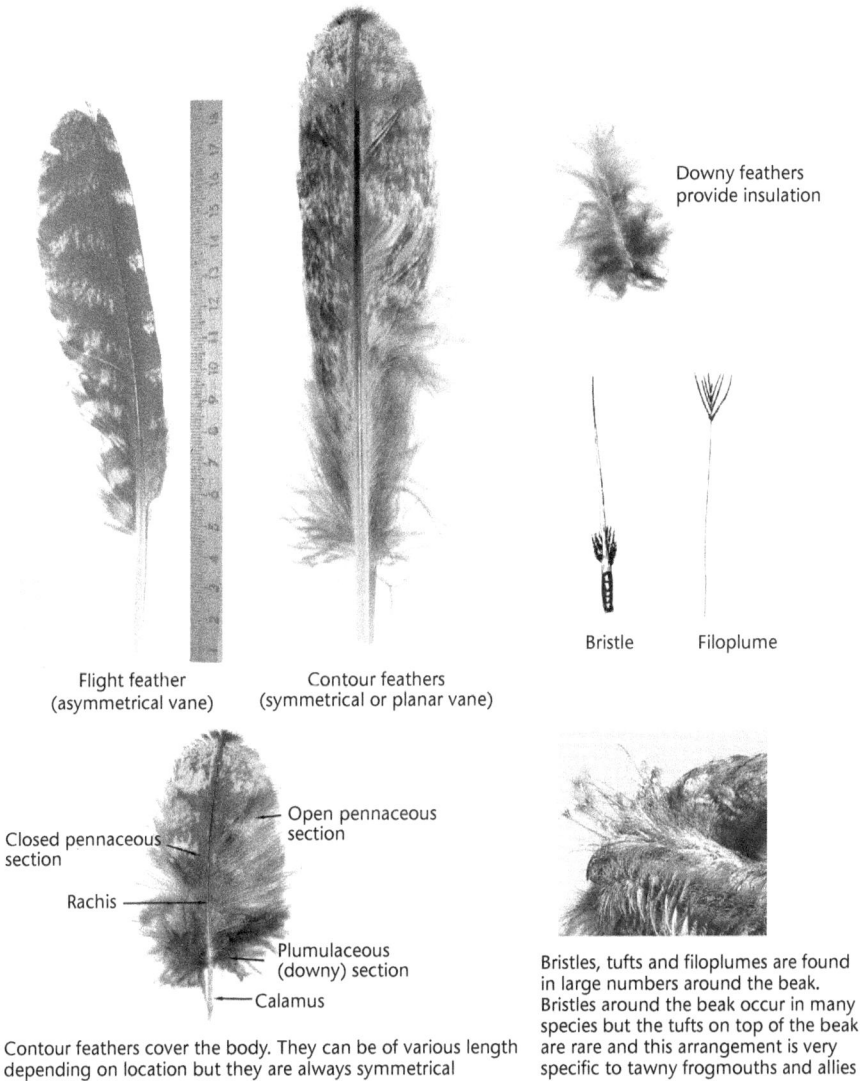

Downy feathers provide insulation

Bristle   Filoplume

Flight feather
(asymmetrical vane)

Contour feathers
(symmetrical or planar vane)

Closed pennaceous section

Open pennaceous section

Rachis

Plumulaceous (downy) section

Calamus

Contour feathers cover the body. They can be of various length depending on location but they are always symmetrical

Bristles, tufts and filoplumes are found in large numbers around the beak. Bristles around the beak occur in many species but the tufts on top of the beak are rare and this arrangement is very specific to tawny frogmouths and allies

Fig. 2.1. Types of tawny frogmouth feathers. 'Pennaceous' refers to feather strands that are interlocked (called barbules) while 'plumulaceous' sections refer to feather structures without such interlocking barbules. Bristles and filoplumes are specialised feathers, as shown.

and none has ever been found, although some species of parasite are too small to see unaided and small numbers could easily avoid detection. There may also be differences in parasite load between rural and urban birds, likely to be greater in urban populations.[3] The fluffy parts near the skin (see Fig. 2.1) also feel sticky and it is extremely difficult to pull these feathers apart to get to the skin (when needing to give an injection, for instance). Even wetting the feathers locally is difficult because of their water-repellent quality. They also form such a thickly matted layer that it protects nestlings from ticks or insects. The entire body is simply so well wrapped in plumage that access for any parasites appears to be particularly difficult. It is, of course, possible that diet and other factors may combine to make the protection against parasites particularly effective. This has not been studied. Noticeably, however, tawny frogmouth nestlings and juveniles seem to have considerably fewer problems with parasites than other altricial species. Altricial species are those species that raise their young in nests, often for several weeks, and they do so because their offspring are completely helpless: they often have their eyes closed for several days post-hatching, cannot feed themselves and often cannot thermoregulate, that is, maintain their body temperature. Many altricial hatchlings have bare skin and would die without the bodily protection of a parent bird (see Chapter 7 for more detail). In case of build-up of faecal matter, problems of parasites and the presence of insects may become increasingly acute, especially for the nestlings.

In addition, the type of feather cover of nestlings and juvenile tawny frogmouths almost completely hides the actual shape of the bird (see Fig. 2.2). The beak is barely visible and there is not even a hint of the eyes when they are closed. Once I observed a tawny frogmouth nestling that had fallen to the ground. Five magpies, considerably larger than the nestling, approached and inspected this odd feathery object. They looked at it, looked at each other, continued to give brief vocalisations to each other as if discussing the case, and after about five minutes staring at the nestling, they leisurely walked away and eventually flew off. It seemed that they had not resolved the phenomenon but were satisfied that it posed no danger and, interestingly, showed no interest in attacking the nestling either. One wonders whether camouflage merely refers to blending in and hiding or whether it not also suggests something more: that the individual, once discovered, is of a shape, colour and kind that does not arouse antagonism. In the case of young tawny frogmouths, it seems clear that the off-white/greyish plumage does not carry a message of danger. It is intriguing that the bird was not attacked by any of the usual warriors of the bush: magpies, ravens, currawongs, butcherbirds, noisy miners, white-winged choughs, magpie larks, indeed, a whole assemblage of species that normally do not hesitate to attack an injured or stray bird. Staying motionless during such inspections (playing dead) may also decrease interest. Perhaps the most important aspect of the effectiveness of the camouflage is that the nestling seemingly keeps its eyes closed during the inspection by other birds, leaving only a slit open (more of this later). Owl nestlings (even when of similar size and colour as tawny frogmouth nestlings) keep them open and they are attacked vigorously.

Filoplumes and bristles are found exclusively on the head of tawny frogmouths (Fig. 2.3). As in nightjars, bristles (single strands) overhang the beak and are shaped forward along the beak, although not as prominently as in nightjars. In nightjars, the bristles are positioned so as to be draped around the beak and they

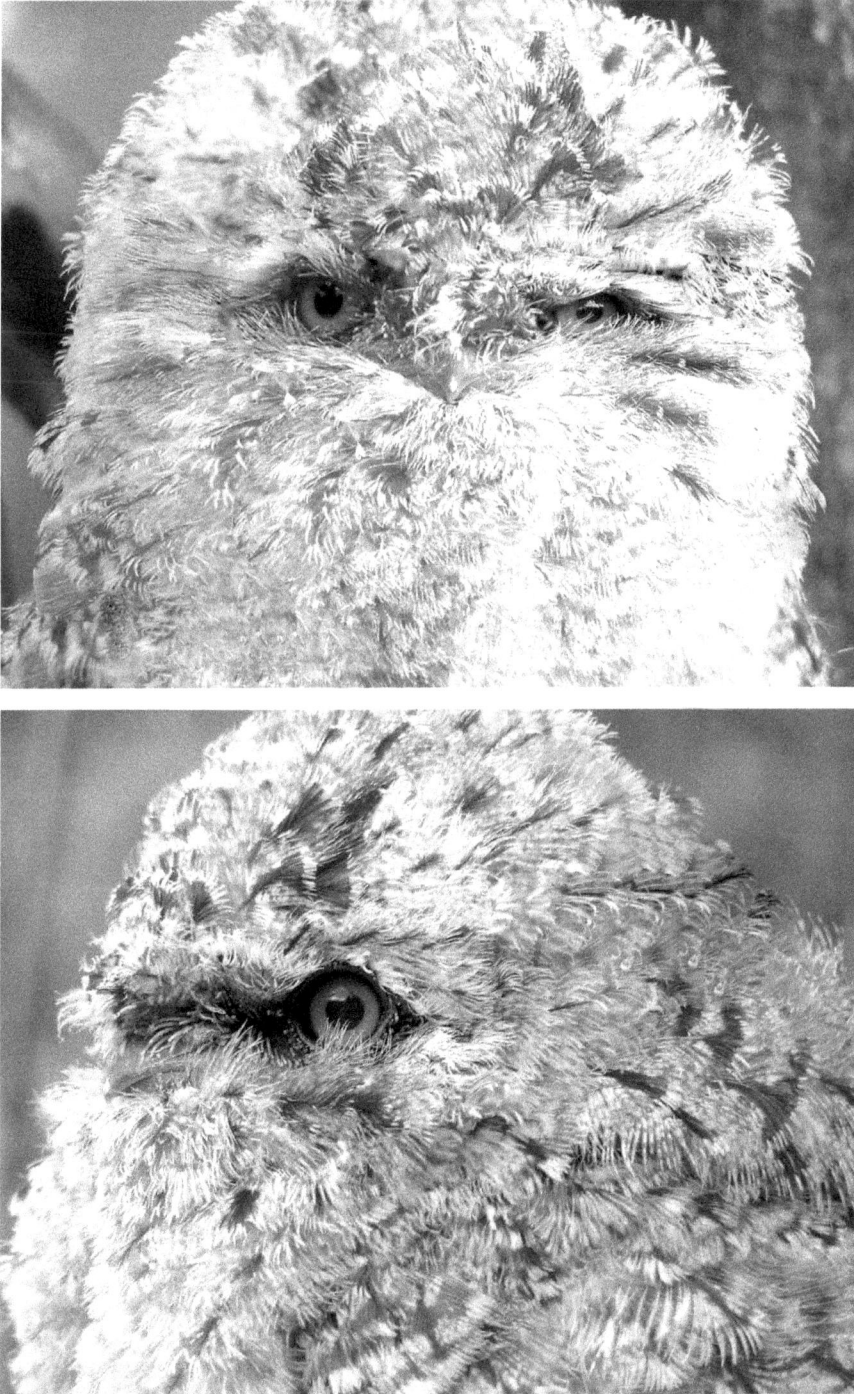

Fig. 2.2. Nestling plumage: frontal (top) and side (bottom) views of the same bird. The beak and eyes are well covered. The frontal view clearly shows that the head and body form one continuous contour.

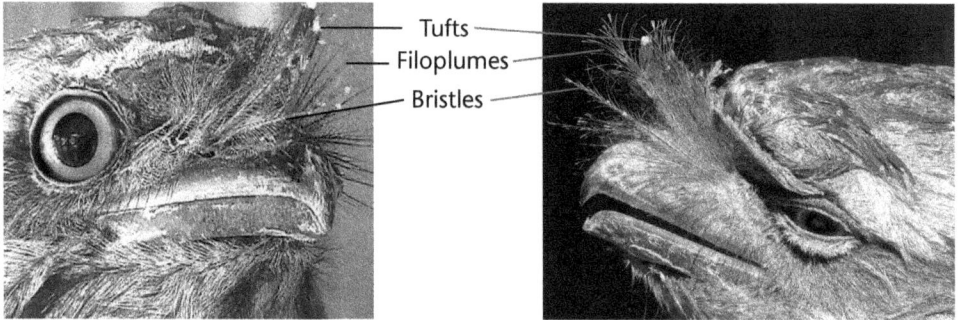

Fig. 2.3. Head plumes, tufts and bristles. These elaborate adornments around the beak could well help to prevent venomous live prey from stinging the bird.

do open out to the front of the beak. It is thought that these bristles serve the purpose of netting in insects during feeding flights. Whether or not this is the case in tawny frogmouths cannot be answered with certainty, even though tawny frogmouths may also catch insects on the wing (although this is not their only method of foraging) (also see Chapter 5).

The bristles above the beak that form a conspicuous tuft are as baffling in structure as in function. They are quite unusual among birds. Although protrusions of the integument, such as combs, wattles and ear lobes are familiar sights in domestic fowl (chickens, turkeys) we know that, in these cases, the protrusions are part of sexual dimorphism and quite often function as sexual signals. The tufts above the beak of a tawny frogmouth, however, are the same in both sexes. It has been suggested that they serve as camouflage by hiding the contours of the beak. That might be so. It is possible to conceive, however, that the copious bristles and tufts around the beak, from the lower mandible to the upper mandible, extending above the nares to the forehead, may serve a far more immediate purpose. Tawny frogmouths (as shall be discussed in more detail later) feed on several venomous invertebrates with mobile bodies and flexible stings. Having watched them feed on centipedes, the wriggling body held in the beak is caught by these tufts and protects the eyes and any part of the skin from being stung.

Tail feathers (rectrices)

The tail feathers (10 in all) can be fully fanned out to take a wedge-shaped form, useful for steering among trees at low speed. When folded, only the two longest feathers usually show, making a fishtail pattern or a shape like that of a folded moth wing. It can be seen in Fig. 2.4a that the tail feathers are of various lengths and design. Most of them taper off to a very pointed tip but outside rectrices are more rounded. Moreover, the centre feathers are entirely symmetrical while those located further outside are asymmetrical as are the primary feathers (Fig. 2.4b).

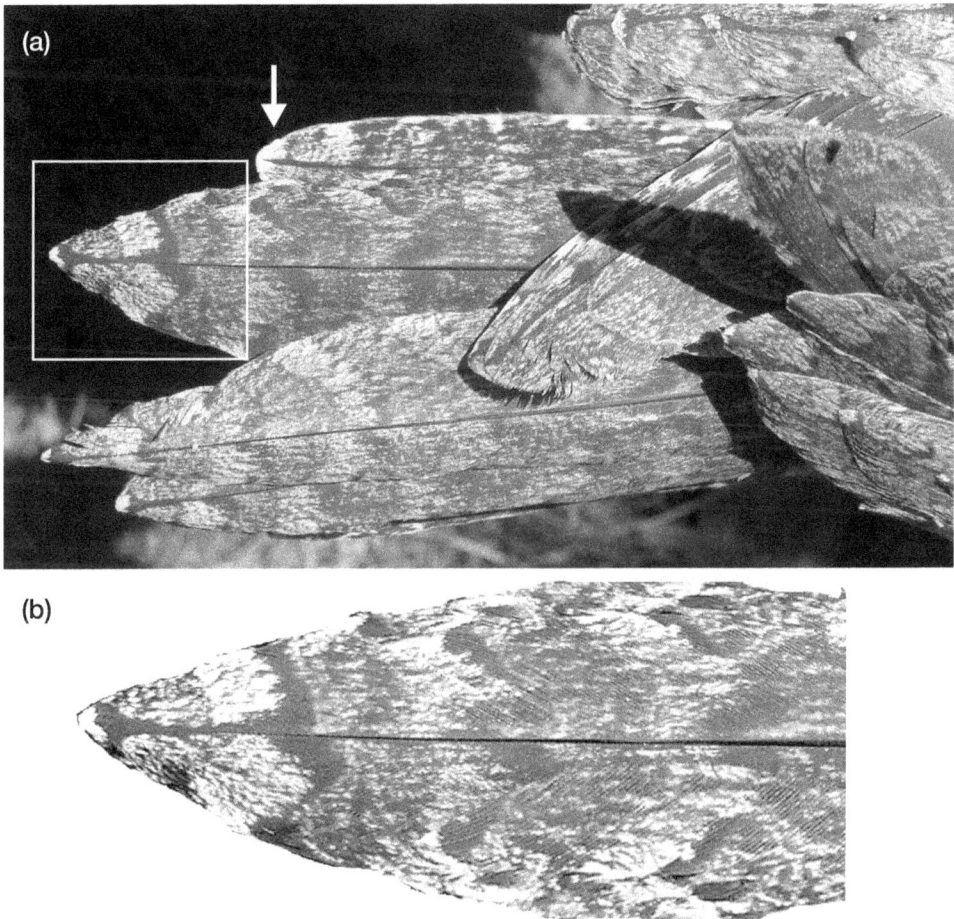

Fig. 2.4. The tail feathers are arranged in pairs and staggered so that the top tail feather pair is usually the shape that can be seen. Note that the central pair (highlighted with a square in (a) and shown enlarged in (b)) is far more tapered than the pairs underneath (arrow in (a)). The sharp tapering makes the tail appear forked.

Moult

To ensure that feathers remain in excellent condition, all birds undergo feather moults, meaning that feathers have to be replaced periodically. This happens throughout the birds' lives. One of the difficulties of moulting is that birds that fly cannot shed too many feathers at once or their flight would be compromised, as indeed happens in some species (ducks for example). Timing and extent of moults differ between species. In tawny frogmouths, adult moults follow the patterns of what is referred to as 'staffelmauser'. Staffelmauser, a German word, meaning literally 'staggered moult', indicates a very specific way of feather replacement. Unlike earlier opinion claiming that tawny frogmouths moult in

Fig. 2.5. The thick down of nestlings, covering even most of the face and beak, offers excellent protection against wind, rain, cold and even heat, as well as camouflage. Feather type and colour changes, even as nestlings and juveniles, are made possible by staggered moulting. This is unusual and in this regard tawny frogmouths are rather exceptional.

December/January[4] it is now thought that tawny frogmouths undergo a gradual change of plumage over time so that it is difficult to identify a specific moulting time or period.[2] Juvenile changes in plumage, by contrast, are easily determined because plumage type, colouration and hue change in distinct periods of development. The initial white down is changed to juvenile down when the birds are still nestlings (Fig. 2.5). Next, flight and tail feathers appear by the third week after hatching. They have adult colouration and are fully pennaceous (meaning they are interlocked, see Fig. 2.1). Around the time of fledging, sometimes slightly before and sometimes slightly after, the first post-juvenile (pre-basic) moult begins, making the plumage look like that of an adult, although the bird is still much smaller than an adult. Any further moults and changes continue as staffelmauser.

Wings

The wings follow the general pattern of wing construction and have two main divisions: the main flight feathers (primaries) are attached to the manus (metacarpal bones and phalanges) and the secondary feathers are attached to the ulna. Flighted birds usually have between nine to 12 primaries.[1] Tawny frogmouths have 11 primary feathers and 12 secondary feathers (this varies between six and 32 across species). The wings are broad and rounded, common to forest flyers (Fig. 2.6).

A special feature of tawny frogmouth wings is found on the underpart of the so-called leading edge (patagium). The patagium is a very strong and flexible tendon that enables the bird to stretch the wing and keep the frontal part of the wing nicely stretched like a sail. This is a remarkable and simple anatomical feature that keeps wings mobile and tense at the same time but, incidentally, proves to be one of the greatest fatal injury risks. Once torn, by barbed wire for instance, it is irreplaceable and irreparable and forever ends the ability for the bird to fly so the bird will eventually die. In diurnal birds, there are no special feathers underwing and that means that, if close enough, one can hear the sound of the wing in flight, from a small hush sound to, sometimes, as in magpies and crested pigeons, a sharp wing hiss. In nocturnal birds, such as owls and tawny frogmouths, the underside of the wing, close to the leading edge, is generously equipped with a special layer of feathers that serves the sole and important purpose of absorbing any noise created by air turbulence (Fig. 2.6). At least to the human ear, tawny frogmouths, just as owls, have entirely silent flight because of this design.

The integument, including the feathers, is also involved in communication. Musculature controls body feathers and head feathers are controlled by an ingenious system located in the subcutaneous area; this allows for some mobility and repositioning.[5] In some species, states of arousal and sexual maturity can be

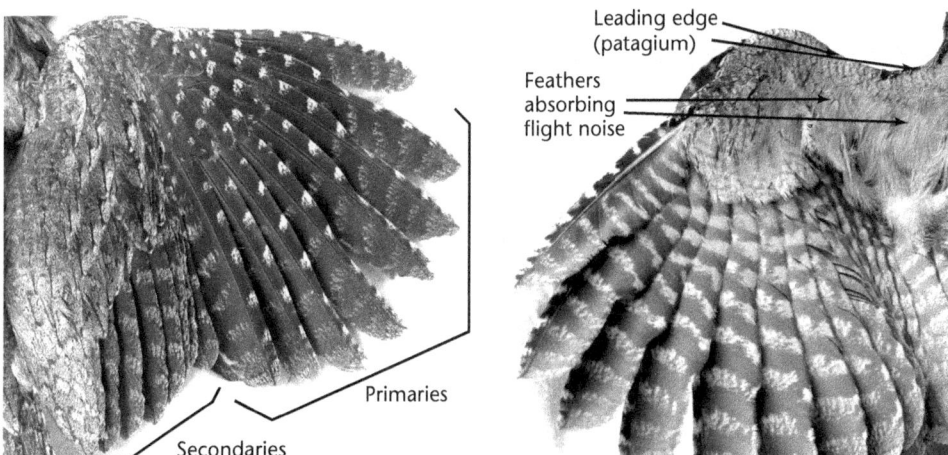

Fig. 2.6. The right wing viewed from above (left) and from below (right).

Fig. 2.7. All the head feathers are extended when caught out in a hiding spot – mild alarm and defensive action to make itself look larger and make the unknown/potential danger go away. This actually works in most cases.

ascertained by extraneous features, such as a turkey's snoods or the chicken's comb. Colour changes indicate sexual maturity or deepening of colour can indicate agonistic states. In other species, such as the sulphur-crested cockatoo, the crest is used to make signals.

In tawny frogmouths, feather positions on the head, be this on the crown, the forehead, cheek, chin and throat and, if less obvious to the human eye, even over the whole body, can have important communicative functions. Indeed, tawny frogmouths, despite their reputation of being stoic and seemingly disinterested in the world, appear to use the feathers to display many different emotional states (Fig. 2.7). They can do so by extending or sleeking down feathers on just about any part of their bodies and are possibly more agile and diverse in such feather expressions than any other avian species (see also Chapter 8).

Eyes, ears and olfactory system

The ear, the olfactory system (discussed in Chapter 3), the eyelashes and the eyelids are also part of the integument. Tawny frogmouths are among the few species of

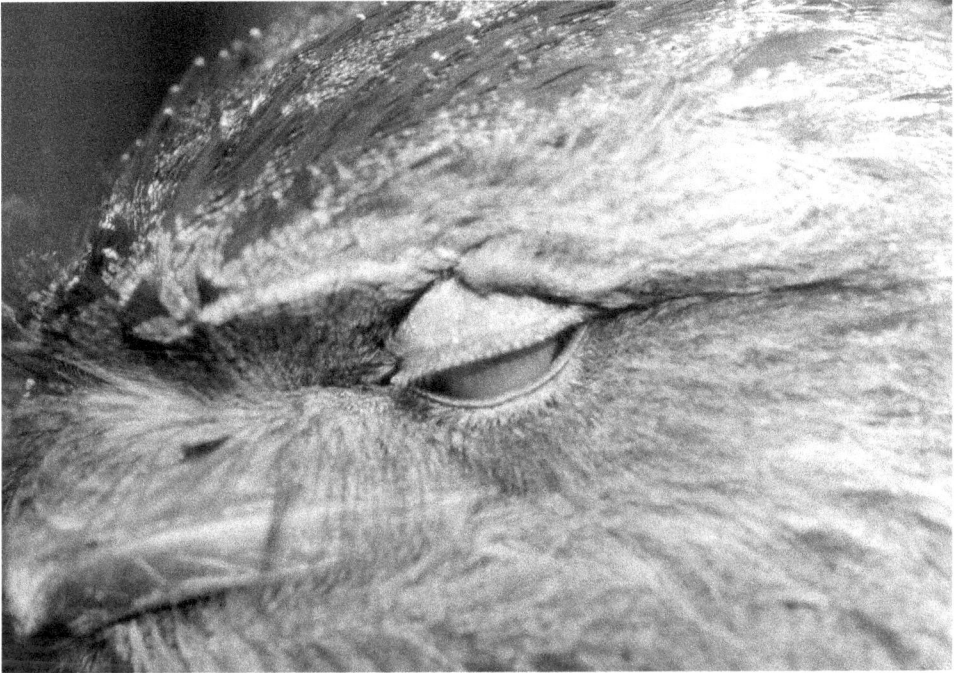

Fig. 2.8. The eyelids of a tawny frogmouth. The nictitating membrane is not transparent as it is in many other birds, as can be seen here (membrane closed across the eye for milliseconds at a time).

birds with pronounced eyelashes (bristle feathers). Also, as in owls, the upper eyelids are thick and designed to close the eye for sleeping. There are no glands in this eyelid.[1]

The so-called 'third eyelid' is known as the 'nictitating membrane' and has a gland that provides fluid to keep the cornea moist and move away any dust particles. It closes in a transverse fashion from the bottom corner of the eye across and upwards.

In most birds, this third eyelid is either completely or almost transparent but in tawny frogmouths (as in only a few other avian species, including owls) it is cloudy (see Fig. 2.8). Presumably, retaining vision even in the few split seconds of closing could be enough risk for a diurnal species but nocturnal species may not need the same level of vigilance[1] or their eyes may not function as well in broad daylight.

Beak

The beak is part of the integument and one of the very pronounced features of the tawny frogmouth. There are ~16 main designs of beak across the class Aves, with a multitude of modifications, and each form is congruent with the feeding niche that a species has evolved to occupy (see next chapter).

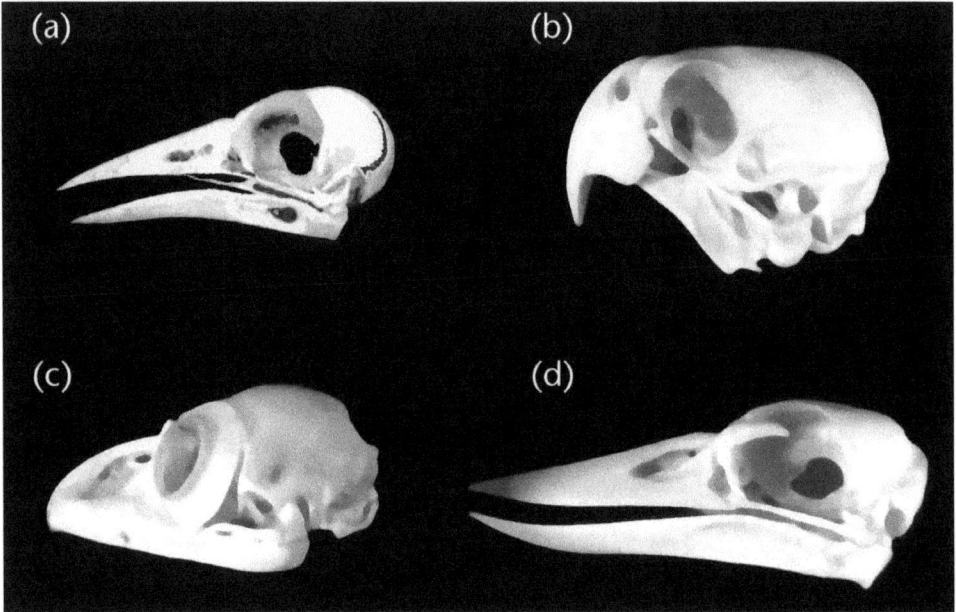

Fig. 2.9. Skulls and beak shape of four large Australian birds: magpie (a), galah (b), tawny frogmouth (c) and kookaburra (d). The tawny frogmouth's beak shape is unique. Adapted with permission from Rogers, L. (1995). *The Development of Brain and Behaviour in the Chicken.* CAB International, p. 185.

In Fig. 2.9 some other beak shapes are shown to illustrate these differences. The lower mandible of the tawny frogmouth has a particularly strong hinge and this structure alone, apart from effective musculature, helps one to understand why this bird has a substantial bite force. The edges of its mandible are sharp, although not as uniformly sharp as those of galahs or kookaburras (also shown in Fig. 2.9), so this beak is perfectly designed to crush and pulp food, thus ideal for crushing ectoskeletal invertebrates (discussed in Chapter 5). The lower mandible is completely flat (apart from the edge) and at the tip of the beak fits tightly just below the curvature of the upper mandible. The slight curvature at the tip of the beak assists in holding food. The beak is usually a dull grey colour and blends in with the plumage.

In tawny frogmouths, the beak relative to body size is far larger than in most other avian species, to the extent that it raises questions about weight distribution and aerodynamics. How can a bird fly effectively despite a beak that seems to make it top-heavy (front load in flight), especially in males? The solution is particularly ingenious. As in pelicans, the lower mandible is not a solid structure but merely a triangular bony frame over which the skin is stretched. All birds have some small section towards the back of the lower mandible covered merely in skin but in tawny frogmouths this is extended to the entire width and length of the lower mandible.

Neck and vertebrae

Birds have more vertebrae in the neck region than mammals, and this gives birds generally a greater degree of mobility of the neck than in mammals. The additional mobility is needed because eye movement tends to be rather limited. Neck mobility compensates for the eyeball's lack of mobility and allows visual scanning of the environment. A mobile neck is also specifically necessary to bend the head back for preening. Vertebrae are subdivided into three segments: *1*, the cervical vertebrae in the area from the head to the first complete set of ribs, followed by *2*, the thoracic vertebrae (with rib cage) and *3*, the lumbar vertebrae, towards the rump, that are fused in tawny frogmouths. The mobility of the neck in a bird that hunts for its food (although, in frogmouths, in the least energy expending way) is particularly important in order to follow a prey item while maintaining it within the field of binocular vision. However, turning the head too far back or to the side could pinch off the carotid at the neck and so interrupt blood supply to the head. Owls have two carotids, one on each side of the neck, allowing the head to tilt 180° to each side without cutting off blood supply. Tawny frogmouths, however, have only one carotid[2] like other birds and that limits their head mobility considerably. This will be discussed more fully in the next chapter.

Legs and feet

The leg consists of several sections. The bone articulating with the hip is the thigh (femur) followed by the knee joint. The lower leg (fibula and tibiotarsus) ends with the ankle (intertarsal joint). Below this is a relatively large section, called the tarsometatarsus, followed by the foot with its three or four toes (digits). Usually, what we see of a bird's leg is only from the ankle joint downwards (the tarsometatarsus) because the thigh is hidden in the body and feathers reach down to the ankle joint. When the bird sits down it thus folds the legs into almost equal halves, shifting the centre of gravity back for balance against a forward head and neck. Unusually, in tawny frogmouths, most of the leg is hidden under feathers, protecting the legs from insect bites that can cause such terrible illness in other birds (avian pox). It may also have a function for thermoregulation.

Tawny frogmouths rarely use their feet for walking. They use them to sit and pounce (so they are called perch-and-pounce sedentary predators) although they can shuffle along on branches and on the ground but will usually only take a few steps. Their lives revolve around two postures: sitting or flying. Walking plays a very marginal role in the life of a tawny frogmouth.

Feet

Foot shape and toe position varies according to species and function. There are about seven main types of feet in birds and most birds have either three or four toes, called digits.[1] The toes, or digits, can be freely moveable or partially or totally fused, or one of them, usually digit I, may be vestigial as in emus (that is, suggesting that it might have been three-toed once but the third has no longer any obvious function). Passerines (perching birds, including songbirds) and all birds of prey have unequally sized digits on what is referred to as an anisodactyl (unequally toed) foot and the one (first) digit facing backwards is readily opposable. In songbirds, the three forward digits are generally placed rather closely together while those of raptors are spaced widely apart. Another type of foot has two digits facing towards the front and two towards the back. This is called a zygodactyl foot (from the Greek 'zygo', meaning yoke and 'dactyl' meaning toed) and is typical for parrots and many cuckoos. There is a variant of this in owls and ospreys, in that they have a zygodactyl foot but they can move the fourth digit from the back position to the front.

The reason for describing these two types of feet in detail is that the tawny frogmouth foot is a mixed form of both types (Fig. 2.10). Although tawny frogmouths have the typical anisodactyl foot common to perching birds, the second digit is so mobile that it can be brought from the forward position almost to the back position or can be positioned to the side. Functionally, in parrots and cockatoos, the pairing of two digits together (two front, two back) is for the purpose of grasping food. Parrots pick up nuts and fruit by the foot and use it as a hand while dissecting the fruit or cracking a nut. Birds of prey have widely separated digits so as to grasp their live prey in such a star-like manner that it cannot escape.

Tawny frogmouths do not use their feet for grasping or holding food so what function does this highly moveable and opposable second digit have? Perching toes are designed to grasp a branch at right angles, as is shown in Fig. 2.10. It has often been observed that tawny frogmouths position their bodies in line with a branch when sitting, rather than at right angles to it. How do they achieve balance then if they cannot take advantage of grasping at right angles to the branch? The answer is provided by the close-up image of a tawny frogmouth roosting aligned with the branch in Fig. 2.11. Here, the function of the second digit is abundantly clear. It is draped around the curvature of the branch in such a way that the second digit, together with the first, provides a balancing stop to the foot while the other two frontal digits are angled away from the second digit by ~45° to reach over the other side of the branch.

Frogmouths have been described as weak-footed or weak-toed presumably because they do not use their feet much for walking. Their ability to balance perfectly comfortably and still on a branch should persuade one otherwise. Their toes and feet are not really 'weak'. Most of the weight of the bird would have to rest on the second digit to achieve balance. This posture would present difficulties in most passerines. Presumably, the structure of the second digit in tawny frogmouths

Fig. 2.10. The foot of a tawny frogmouth with the four digits in the relaxed posture: three forward and one behind.

is an adaptation to the type of camouflage they adopt (as will be described in detail in Chapter 4).

Note that the entire foot is covered by near equal segments. These scales are similar to those of other birds although the segments are usually not so pronounced. In tawny frogmouths, the scales are grey in colour. In Fig. 2.10, the fourth digit shows particularly clearly the separate pad under each segment: nine pads for the fourth digit alone. The padding just behind the claw is the most pronounced.

Such detailed anatomical attention to padding is an adaptation to the amount of perching tawny frogmouths do. Apart from kookaburras, tawny frogmouths are probably the champions among perchers, spending literally most of their lives sitting and perching. Despite this, their feet look as good as new and tend to be in excellent condition through their entire life. Not many species can boast such

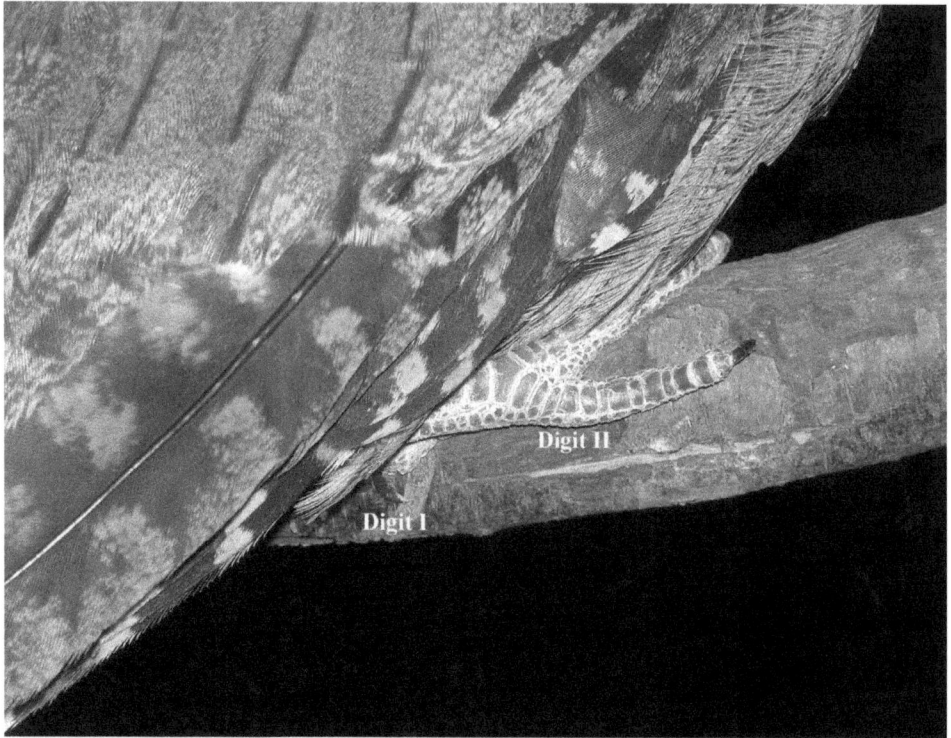

Fig. 2.11. The foot of a tawny frogmouth parallel to the branch during roosting. Note that the second digit is splayed out around the branch and this digit actually holds a good deal of the weight of the bird.

healthy feet, particularly with such constant and unchanging use. This is surely a brilliant design feature in tawny frogmouths, showing that the earlier description of frogmouth feet as 'weak' is not just misleading but factually incorrect.

3

The brain and the senses

Although much is known about perception in birds, relatively few species have been studied in detail. As for the tawny frogmouth, little is known about its hearing, senses of smell or taste and tactile sense. One major study of the tawny frogmouth's sense of vision is all we have.[1] In general, hearing and vision are likely to be the most important senses for the majority of birds, although there are surprising adaptations that have made one of the less important senses of paramount importance for some species (for example, the sense of smell in petrels and tactile sense in kiwis). It has often been said that nocturnal birds rely more on their sense of hearing than vision in order to negotiate obstacles in a dark world and find food, raise offspring and escape predators and injury. Clearly, however, the eyes of nocturnal birds are specially designed to capture as much light as possible and to allow for assessment of distance. Since information from the various senses is coordinated and interpreted in the brain, it will be discussed first.

The brain

In recent years, avian cognition has developed into a separate field, and with considerable input from neuroethology, has explored the avian brain, identifying forebrain nuclei that may contribute to avian cognition in certain areas. Tawny frogmouths have rarely been thought of as birds worth investigating for their ability to do anything cognitively remarkable. Interestingly, however, the uncertainty as to where they belong taxonomically – whether to nightjars or to owls – has led to questions as to how well the tawny frogmouth is equipped for nocturnal life and how its behaviour and sensory abilities compare to owls on the one hand and to nightjars on the other. Sensory perception in nocturnal birds has to be adjusted for low light conditions, and perhaps it needs to be specifically equipped for detecting motion and also perhaps for auditory perception. We know that barn owls can pinpoint moving prey items such as mice in complete darkness, relying on audition alone. Tawny frogmouths prefer to hunt when they have some light but they can equally hunt in the dark (more of this later). We also know that tawny frogmouths have rather large brains relative to their body size.

It has been quite commonplace to grade the cognition of birds according to their overall brain volume adjusted for body size. However, in tawny frogmouths, interest in brain size was not motivated so much by discovering its cognitive abilities but to clarify its taxonomic position between nightjars and owls.

Indeed, based on the ratio of brain volume to body mass, the tawny frogmouth seems to be in a slightly unusual position. The tawny frogmouth is one of the largest frogmouths in the world so one would expect a larger brain size than in the smaller birds but, adjusted for bodyweight, the tawny frogmouth also seems to have one of the largest brains among its nightjar and frogmouth relatives and a brain even larger than that of the oilbird, a bird in the same weight range (Fig. 3.1a).

If it is argued that an exceptional brain volume to bodyweight ratio in addition to a range of physical features and lifestyle of the tawny frogmouth, adds weight to the argument that tawny frogmouths ought to belong to or be shown to derive from owls, then the ratio of brain weight to body mass would seemingly speak against this (see Fig. 3.1b).

Iwaniuk and colleagues[2] have argued for the relatedness of owls and tawny frogmouths not on the basis of overall volume but on the basis of just one element of the brain, the Wulst. The Wulst, now called the hyperpallium, is a specific conglomerate of neurons that is part of the forebrain. They argued that, on the basis of the relative size of the Wulst in the caprimulgiform birds, both the oilbird and the tawny frogmouth should be placed outside the caprimulgiform (nightjar) group. As they rightly argue, this does not resolve the issue but brings into the debate yet another important variable.

The Wulst is related to vision. In raptors such as eagles, hawks and owls, it is large and well developed and species with more frontal orbits and broader binocular fields, such as owls, have relatively large Wulst volumes. It has been very well researched in a variety of species that the visual Wulst neurons are selective for orientation, movement direction, spatial frequency and binocular disparity use. The larger the binocular field, the greater the size of the Wulst. This refers to orbit orientation of the eye, i.e. the ability of the bird to focus both eyes in parallel and to the front. Hence, the Wulst is largely related to binocular vision and at times also to stereopsis, i.e. one of the ways the brain enables depth perception.

There is also a portion of the Wulst that may process information of touch, in this case from the feet. This is very important in birds of prey. They view their prey

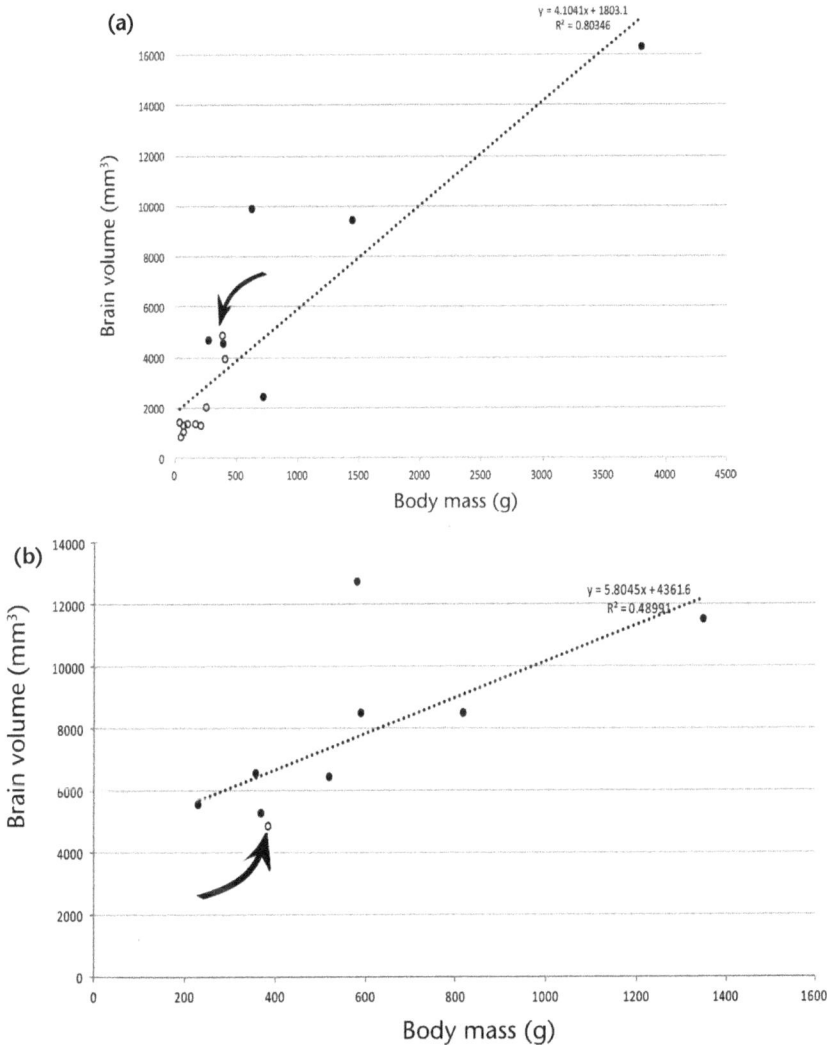

Fig. 3.1. Relationship between brain volume and body mass in the tawny frogmouth and other birds. (a) The tawny frogmouth (arrow) is compared to a representative group of nightjars (hollow circles) and to several other large sized native Australian birds from different orders (black filled dots). The weights of these birds (shown in brackets) are: Australian magpie (280 g), Australian raven (635 g), sulphur crested cockatoo (726 g), brown goshawk (403 g), osprey (1450 g) and wedge-tailed eagle (3814 g). Note that there is a cluster of eight nightjars in the weight range of 41 g to 257 g, all with relatively similar brain/bodyweight ratios (from lightest to heaviest): the Australian owlet-nightjar, whip-poor-will, spotted nightjar, large tailed nightjar, chuck will's widow, white-throated nightjar, Nacunda nighthawk, and the grey potoo. The oilbird (414 g) is above the nightjar trend line and the tawny frogmouth (387 g, see arrow) is well above all other nightjars. (b) The tawny frogmouth (marked with arrow, hollow circle) is compared with Australian owls (black circles) of two families (Ninox and Tyto). The weights of these birds vary quite substantially (shown in brackets, as averages): the powerful owl (1350 g), rufous owl (820 g) and masked owl (593 g), barking owl (592 g) and sooty owl (585 g), barn owl (360 g) and grass owl (372 g) and the southern boobook owl (234 g). In weight range, the tawny frogmouth (388 g) fits in well, being slightly smaller than the barking and sooty owls and slightly larger than the barn and grass owls, but owls in general have larger brain/bodyweight ratios than the tawny frogmouth.

with their eyes and shortly before capture they extend their legs and talons and only then capture the prey item. Hence, foot and eye coordination is of utmost importance for raptors because the event of the capture relies on absolute precision of making contact with and firmly gripping the intended prey item, an event that is measured in milliseconds. In all birds of prey, including owls, it is the Wulst that coordinates these events. Fig. 3.2 shows the position and superficial appearance of the Wulst in several species. The extraordinary coordination between eyes and feet via special areas of the Wulst was recently discussed in barn owls, showing that a part of the Wulst (the rostral somatosensory part) forms a unique physical protuberance dedicated specifically to the representation of the contralateral talons.[3]

While the connection between the visual and the somatosensory Wulst explains very well how capture of a moving prey is coordinated and why the Wulst is so large in birds of prey, it may not serve exactly the same explanatory purpose in tawny frogmouths.

In tawny frogmouths eye/foot coordination is not necessary for the capture of their prey. As was discussed in detail in the previous chapter, tawny frogmouths do not use their feet at all for capturing or manipulating prey. It is all done exclusively with the beak. Unlike owls and other birds of prey, frogmouths do not possess talons with long sharp curved claws – a specific design feature in birds of prey to pierce even strong hides and firmly lock their talons once the prey is captured.

While the volume of the Wulst relative to brain size may be seen as an indication of the similarity between tawny frogmouths and owls, it is worth noting that other entirely unrelated species, such as Australian magpies, also have a relatively large Wulst (Fig. 3.2). Magpies are songbirds, not birds of prey, and they are not even distantly related to tawny frogmouths. However, magpies can use their feet to manipulate food items, just as cockatoos do and most parrots can.

The size of the Wulst is thus a very specific biological adaptation to the needs for hunting and prey capture in birds and it contributes perhaps even to a greater overall brain weight or, alternatively, the brain has reserved a relatively large area for one specific and essential function without any significant development in other areas. The role of vision and audition in crepuscular feeders and in birds of the night has fascinated researchers for a long time. The barn owl is probably the best studied species in this regard but some attention has also been paid to the tawny frogmouth.

Vision

Frogmouths and owls have probably the most unusual arrangement of the eyes of any bird species. Tawny frogmouths have dramatically large eyes with a structure that allows the bird to see under very dim light conditions but still enables them to see to fly during the day. Most birds, including birds of prey, have eyes that are laterally placed, less so in birds of prey than in songbirds. Owls and frogmouths have their eyes positioned frontally (Fig. 3.3).

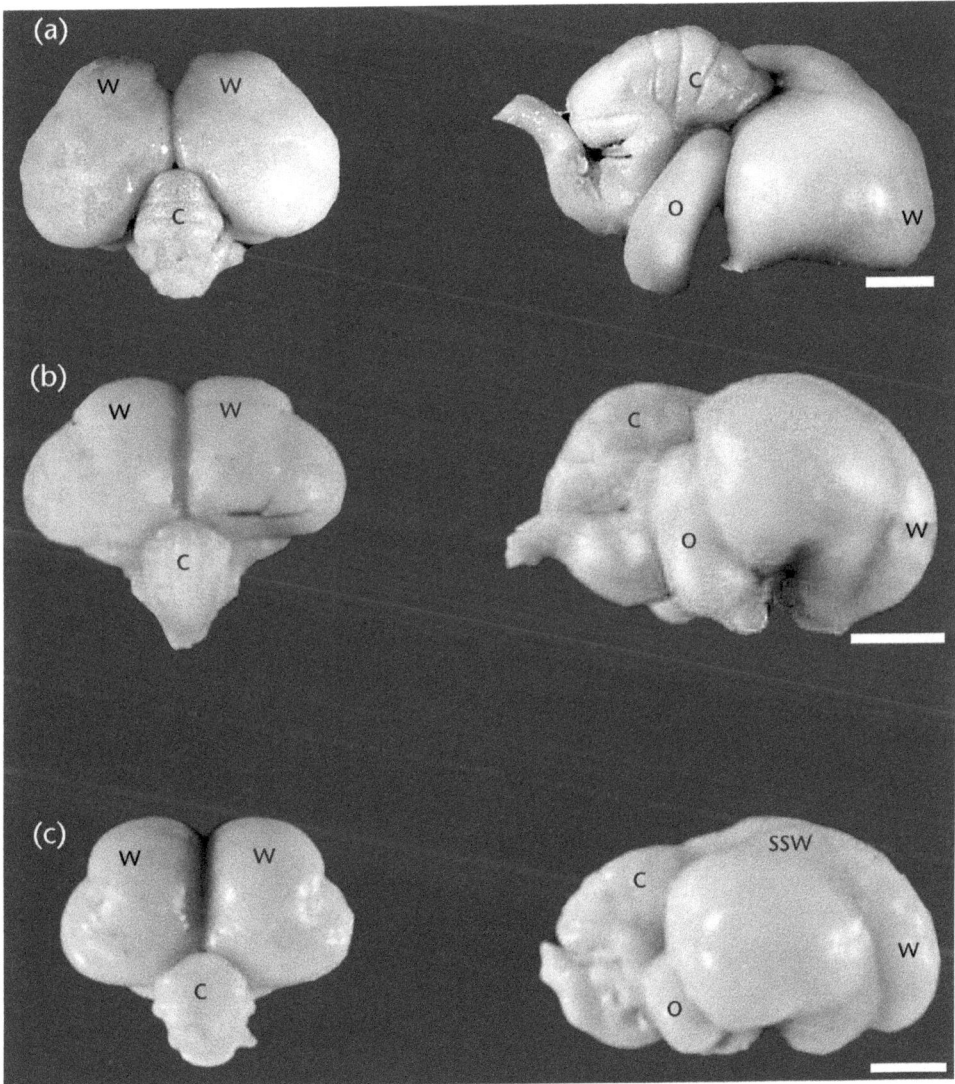

Fig. 3.2. Comparison of the brain and especially the Wulst (hyperpallium) area in (a) the Australian magpie, (b) the tawny frogmouth and (c) the barn owl. In the first vertical column the brains are viewed from behind (dorsal/caudal); in the second column they are viewed from the right side (laterally). Some key areas of the brain are marked in all three birds: w, Wulst (note that it is located at both sides of the brain); c, cerebellum (part of the hindbrain); o, optic tectum. In the lateral view of the barn owl brain (c), the Wulst is marked twice: w for the general visual Wulst, and in the same general area is the protrusion of ssw, the somatosensory Wulst, enabling fast eye/foot coordination. The figure shows that, in all three species, the area of the Wulst is a visible protrusion, most clearly distinguishable in the barn owl (c) but also clearly evident in the tawny frogmouth (b).

Creating a binocular field of vision

An image is projected onto the retina of the eye using the cornea and then the lens. The retina at the back of the eye has special nerve cells, called photoreceptors, which transform the light energy into electrical energy and transmit this along the optic nerve to the brain where the visual information is processed. The cornea and

Fig. 3.3. Frontal position of the eyes in nocturnal hunters (top row, from left: tawny frogmouth, barn owl, southern boobook owl) and intermediate position in day hunters/diurnal raptors (bottom row, from left: Australian hobby, black-shouldered kite) as compared to the usual lateral position (bottom right: white-headed pigeon).

lens together focus the image on the retina. Each eye can be focussed independently of the other (as tested in chickens[4]) and each eye can be moved independently of the other. Hence, not only can each be focussed at a different distance but each eye can send a different image to the brain.

Visual perception can be monocular or binocular, with large or small blind areas behind the bird's head. Vulnerability to becoming prey is related to the size and type of the visual field. The reason why most birds have their eyes placed laterally is to reduce blind areas and increase the field of vision, as is seen, for instance, in the cattle egret (Fig. 3.4). Clearly, this aids their survival because the eye position allows them to see above, and behind them, a necessity even in flight when aerial predators may single them out for attack. By contrast, wedge-tailed eagles, at the top of the food chain, have less need to look panoramically for the sake of vigilance. Hence, they can maximise the overlap of the visual field and enhance their depth perception (Fig. 3.4).

Thus, there are trade-offs in achieving a large field of monocular or binocular vision; they cannot exist together. Birds that may become prey have sacrificed binocular vision to the monocular field (for greater panoramic viewing) and birds of prey have decreased their monocular vision (with concomitant increase in blind areas), in order to have a greater binocular field of vision. The binocular field of vision is specialised for detailed vision and to make accurate assessments of the distance of objects, known as depth perception,[5] that is vitally important for capturing prey. This is the case in tawny frogmouths. In owls and in tawny frogmouths the binocular field is larger than in any other avian species (almost twice the size). Both raptors, including owls, and tawny frogmouths therefore have

Vertical field   Horizontal field

Cattle Egret

Wedge-tailed Eagle

Tawny Frogmouth

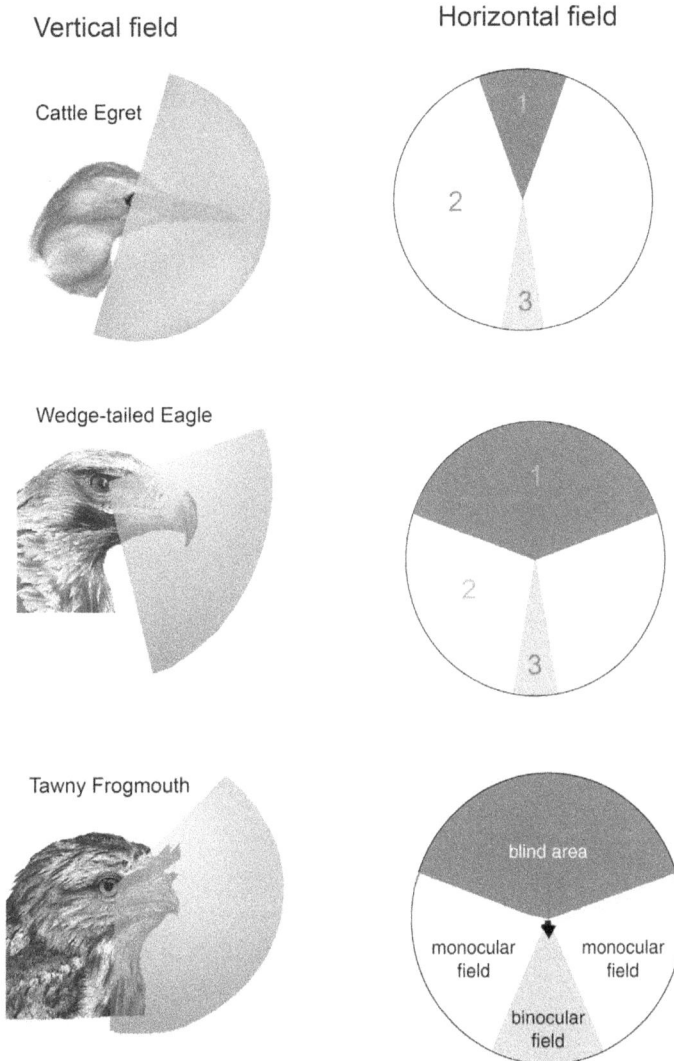

Fig. 3.4. Visual fields of different species. The top image of the cattle egret is a typical visual field for most birds (with some variations). The binocular field in front of the eyes (3) is small but so is the blind area behind the head (1) important for detection of predators. The wedge-tailed eagle, on the top of the food chain, has a much larger blind area and even a smaller range of vision in the vertical field than most birds. The tawny frogmouth belongs to the small range of birds (largely nocturnal) with a very wide binocular field, sacrificing vision in the monocular field (2).

larger fields of blind areas (behind their heads) but only a small field of monocular vision (Fig. 3.4).

Depth perception must be important for tawny frogmouths and will therefore be described in more detail here. There have been some good studies trying to fully understand the importance of depth perception. Depth perception can be achieved in three ways. Tawny frogmouths are able to use two of those, based on observed behaviour, and maybe even all three. One is called 'parallax' (moving the head

Diverged Eyes   Binocular (parallel) viewing   Converged Eyes

Fig. 3.5. The three eye positions of tawny frogmouths: diverged eyes are the relaxed and natural position; the binocular position requires some adjustment in order to position them in parallel; and converged eyes allow for vision below and close-up (but this eye position is usually only held for brief periods).

rather than the eyes), another is convergence of the eyes,[6] and the third is called 'stereopsis'.[7] There is plenty of evidence that tawny frogmouths, especially juveniles, move their heads in a vigorous circular motion (parallax) and that tawny frogmouths of any age can converge and diverge their eyes (Fig. 3.5). Stereopsis is not observable by onlookers. It refers to the visual image seen by each eye in slightly different ways and this image discrepancy is used by the brain to calculate depth. Since tawny frogmouths have frontally placed eyes they may be able to use stereopsis but this has not yet been tested. All owls are believed to have good stereopsis.[8]

Depth perception can be achieved by motion parallax alone. When stationary, an animal can acquire parallax information by moving its head in semi-circular rotation around an imaginary y-axis through the middle of the head, noting that distant objects move less than closer ones as the bird moves.[9] Or an animal can head-cock, which is the fixation of an object in the binocular visual field while tilting the head to various degrees. Head-cocking has been described in a variety of primates,[10] and in birds such as owls and nightjars.[11]

Tawny frogmouths do not head-cock by tilting the head from side to side, but they achieve the same effect by moving the head in a circle or ellipse (motion parallax) while they remain looking in the same direction. Parallax is rarely used by adult tawny frogmouths but tawny frogmouth nestlings and juveniles use motion parallax repeatedly and extensively and they do so not just during the hours of the night but also during the day. The head is moved so vigorously and rhythmically, it looks like a rock dance movement and its physiological purpose may be lost on the onlooker. Indeed, this behaviour may hint at important developmental stages.[12] Such rotation of the head in tawny frogmouths has not been reported previously. It may also suggest that their day vision is relatively good.

Seeing depth (in addition to parallax) by convergence relies on the degree to which the eyes can be converged (turned inwards) to focus on an object. This convergence is very interesting because a detailed study found that the eyes of tawny frogmouths in the resting position are naturally diverged slightly outwards keeping the most sensitive areas of the retina, the foveas, looking somewhat sidewards.[1] Hence, to move the eyes inwards and to focus on an object is a specific action that the birds can use for the degree of convergence needed to determine depth.

In frogmouths, the eyes can move in opposite directions and this has been described as unusual in birds (Fig. 3.5). In most avian species, the eyes can perform small, staccato movements, known as 'saccades'. Both eyes perform such movements simultaneously and in parallel and they do so for scanning. This is well known. However, when tawny frogmouths do this, their eyes do not both move in the same direction but in opposite directions, even simultaneously! Whatever the method, it creates a large binocular field.

The advantage of a large binocular field is related to precision of flying and ultimately also to catching prey. During flight, the visual images flow outwards and backwards through the bird's visual fields. By having the eyes directed forward so that the binocular overlap is large, the optical flow seen by each eye is more symmetrical than in species with laterally positioned eyes. Hence, both eyes see almost the same flow field.[5] This is especially important to nocturnal flyers such as tawny frogmouths and owls. It is unlikely that owls evolved their large binocular field solely for depth perception, especially since we know that some owl species can locate their prey using sound rather than vision.[13] Having two eyes seeing almost the same optical flow might, therefore, be especially important in nocturnal birds with their large eyes designed to collect as much light as possible and a symmetrical optical flow seen by each eye may also help negotiating tree trunks and branches.

Structure of the eye of the tawny frogmouth

The eye of the tawny and other frogmouths has some unique design features. There are three different shapes for eyeballs: elliptical, round and tubular. Tawny frogmouths and, apparently, most owls have tubular eyes.[14] The eye of the tawny frogmouth is possibly the largest eye of any bird in relation to the size of its head. In fact, it fills more than 30 per cent of the skull. The complete eye in a tawny frogmouth can best be seen using X-ray (Fig. 3.6). The X-ray reveals clearly that the shape is indeed tubular and that the major part of the eye is, in fact, hidden inside the skull. Structurally, this could pose a problem but, ingeniously, this has been solved by a bony ring around the eye-socket. It semi-divides the eye into an outer and inner part and firmly holds the visible and narrow part of the eye in place (Fig. 3.7). Imagine a balloon and placing one's hands around it near the middle and squeezing it until two parts emerge while still having a connection between the two parts. It is the same principle. In a few extreme accident cases (a bird's head had been hit hard by a car) a significant part of the eye ball extruded from the socket but it had not broken the eye. We were able to reinstate the eye in its proper position by positioning the eye in such a way that it could be pushed back through the eye ring. Amazingly, later visual tests revealed that the birds had completely normal eyesight.

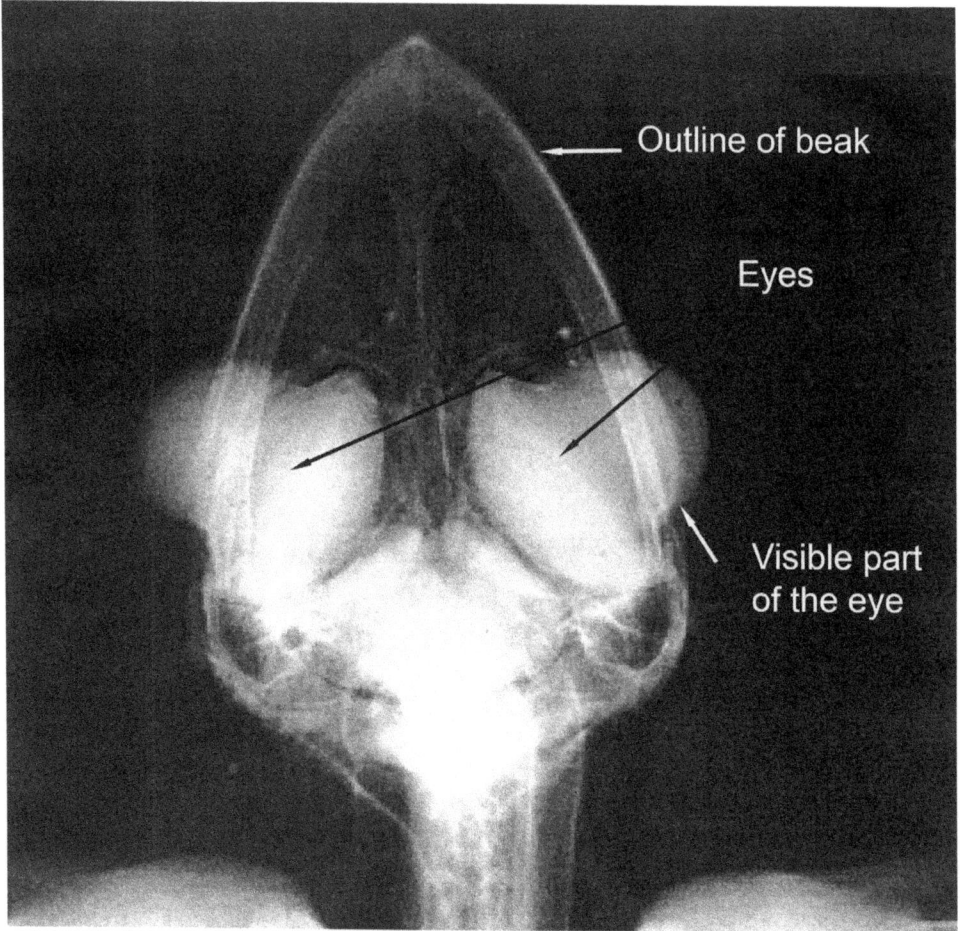

Fig. 3.6. X-ray view of the eyes of the tawny frogmouth. Note the tubular structure that is also typical of many owls. Most of the eye is hidden inside the head and can only been seen using X-ray.

There are some basic advantages and disadvantages to having large eyes. Larger eyes collect more light than smaller ones.[15] The photoreceptor cells in nocturnal birds are spaced widely apart. The image seen by a nocturnal bird is 'grainy' rather than sharp, as in diurnal birds. Yet with larger eyes tawny frogmouths can see better in low light intensities than can birds with small eyes. However, large eyes cannot be moved as easily as smaller eyes can and, therefore, tawny frogmouths need to turn their heads more than diurnal birds. Indeed, the eye socket has the disadvantage that it further limits eye movement.

This does not mean that they have no eye movements. In fact, they have very important ones as already mentioned, such as frequent convergent saccades that turn their eyes inward to look in front[1] and independent eye movements as was shown previously in Fig. 3.5. Head movements are essential though to

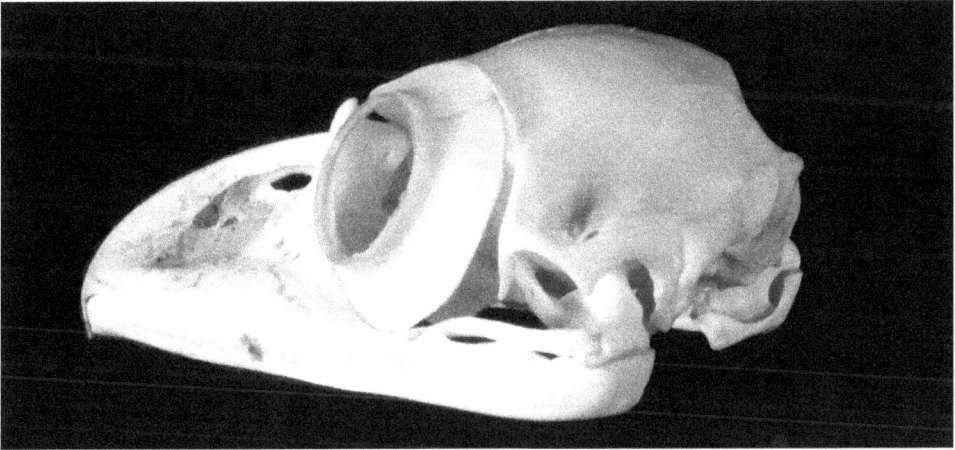

Fig. 3.7. The eye socket holds the large eye in place.

compensate for the inability to make large eye movements and to maintain a
good field of vision.

Pupil and iris

Birds that have to forage for their food in darkness are able to dilate their pupils
widely in order to collect as much light as possible. This is seen in nocturnal
species, such as owls and nightjars, and also in species that enter dark
environments in search of prey. The iris varies in colour in different frogmouths.
Yellow, orange and red irises of different shades and hues have been found in
different subspecies but there are also variations within subspecies.

Surprisingly, I found that tawny frogmouths can use the iris in communication.
A change in size of the pupil can indicate different states of emotional arousal,[16] but
changes in the size of the pupil (indicating something other than mere changes in
light condition) can be found in many species. What is special in the tawny
frogmouth is that the colour of its iris can change and this change is not even very
subtle. It may change from a yellow colour, when it is in a relaxed state, to a reddish
colour when the bird is aroused or distressed. The red colour is strongest around the
edges of the iris and may be caused by dilating the artery, known as the *circulus
arteriosus iridicus*, which circles the iris in this outer region.[15] I have been able to
test this on several occasions. The male tawny frogmouth shows signs of antagonism
every time another frogmouth, male or female, is introduced into its domain. Before
it raises its hackles (see Chapter 8), the outer region of the iris changes from a
darkish yellow to a red in a matter of seconds. Because such colour changes were
invariably followed by raising of hackles, it is likely that the red colouration reflects
a state of arousal indicating anger or mounting displeasure. Humans, of course, give
signals via their pupils rather than the iris and experiments have shown that

dilation and constriction of pupils are used as definite, but subconscious, signals in humans.[17] It is conceivable that the change in colour in the iris of tawny frogmouths may have a signal function.

Colour vision

Nocturnal species make use of photoreceptor cells called rod cells that are specifically responsive to dim light to detect movement but not detail or colour, while photoreceptors called cone cells see colour and detail. All birds have both rods and cones, as do mammals, but nocturnal species have more rods and fewer cones than diurnal species. The black skimmer (*Rynchops niger*), a species that feeds mostly at night, has five times more rods than cones.[18] Owls have even more rods relative to cones than the skimmer. Hence, nocturnal species like the tawny frogmouth sacrifice some detail of colour vision for superior ability to see at low light intensities.

Hearing

We know that the sense of hearing is likely to be of great importance to tawny frogmouths as it is to most other avian species and to all nocturnal species. Moreover, tawny frogmouths vocalise a great deal, often at low amplitude, and part of that vocalisation is for communicating with other tawny frogmouths (see Chapter 8). Yet we know next to nothing about the hearing ability of tawny frogmouths, how important it might be for locating food, and to what distance they may need to detect the calls of other tawny frogmouths. In general terms, we know something at least of the general structure of the ear because of many studies conducted on different avian species, although there are also very important differences between species. Birds in general have no external ear (or 'pinna') to collect sounds and to locate their source. Birds, like all vertebrates, have two laterally placed ears (they have what is called a 'binaural receiver system'), indicating that each ear collects slightly different sounds.

Moreover, in most birds, except some large flightless birds such as emus, the entrance to the ear is covered by feathers. In some owl species the ear is further protected by a skin-flap.[14] Based on my autopsy of a juvenile, it appears that the external ear in tawny frogmouths is also protected by such a skin-flap, although it is not clear whether this skin-flap remains into adulthood (Fig. 3.8). In some owl species, this skin-flap can be raised to enable the bird to hear sounds from behind. It is not known whether this kind of muscular control is available to tawny frogmouths. The external ear can be of various shapes: square, round and vertical. In tawny frogmouths, as I discovered during an autopsy, the external ear is a vertical slit, as is also found in some owl species. From the external ear sound travels to the middle ear and this part of the ear can vary in size. Researchers have found that size is related to function. Species with good to excellent hearing (such as birds of prey and owls) tend to have more spacious middle ear cavities.[19]

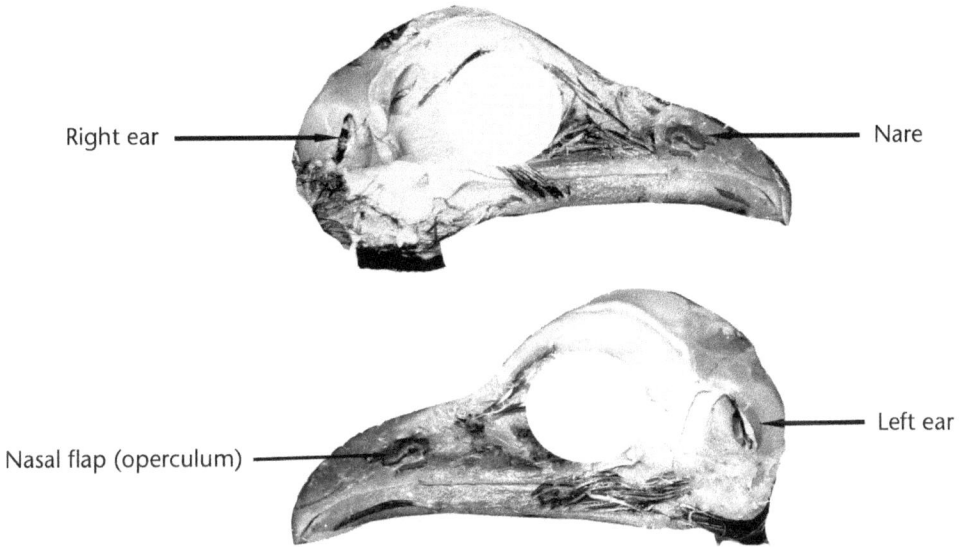

Fig. 3.8. The position of the ears. This juvenile skull and beak have not yet fully acquired adult shape. Note that the ear is positioned as a long vertical slit, slightly tilted, behind the eyes. At this age, the ear is covered by a skin-flap that was removed for the sake of taking these images. (Images from the autopsy of a nestling close to fledging that died as a consequence of a cat attack.)

How well birds can hear has been well tested in owls and this research may be relevant to tawny frogmouths. Roger Payne's classic studies demonstrated the hunting ability of barn owls by providing food under different levels of light, gradually dimming the light to such an extent that food could be located by auditory cues alone. The owl captured the prey accurately every time,[20] although this may also be related to spatial memory. Among birds tested for auditory sensitivity, owls can hear up to 300-fold better than pigeons. The difference may be between nocturnal and diurnal species rather than specifically between owls and other birds. In his book *Birds by Night*, Graham Martin[13] suggests that this hearing ability in owls probably represents the absolute limit of auditory sensitivity in vertebrates. Tawny frogmouths are likely to have excellent hearing, perhaps even similar to that of owls, but this is yet to be verified.

Localising sound

The question is perhaps not so much whether a nocturnal bird can hear its prey but how it manages to localise the sound source accurately. One explanation lies in the very specialised structure of the ear, particularly the middle ear, and another in the way in which some nocturnal birds of prey collect sound. Some owls have facial disks (called ruffs), which are known to enhance an owl's ability to determine the directionality of sound. The disk acts as a sound collector and reflector and for hearing very faint sounds. Such a disk is not only found in owls but also in other birds of prey, such as the pallid harrier (*Circus macrourus*), the marsh harrier (*Circus aeruginosus*) and the Eurasian nightjar (*Caprimulgus europaeus*). Tawny

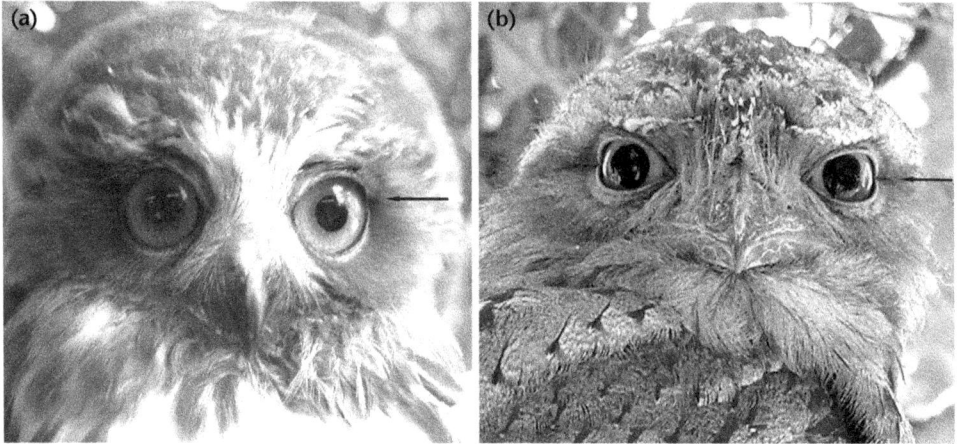

Fig. 3.9. Neither the boobook owl (a) nor the tawny frogmouth (b) have facial disks, as do some owl species, but note that at the midpoint of the eyes the feathers form a groove that might well exist specifically to channel sounds to the ears right behind the eyes.

frogmouths have no distinct facial disk. Yet feathers around the beak are organised in such a way that there is very little difference between, say, the facial feather arrangements of a southern boobook owl (*Ninox novaeseelandiae*) and those of the tawny frogmouth. The feathers around the eyes are arranged in a circular fashion in both species (Fig. 3.9). While this is not quite the same as a facial disk, it seems likely (although has not been tested) that this follows the same principle. Sound may be captured in this large frontal area of the head and funnelled in the direction of the ears directly behind the outer area of these facial feathers. Tawny frogmouths have an additional feature: a feather indentation located near the midpoint of the eyes (Fig. 3.9). This may well function like a channel to direct the sound to the ears located directly behind.

Owls can turn their heads by nearly 180° to either side and by then using a disk-like sweeping motion (similar to parallax in vision), they can locate the source of sounds very accurately. As said before, because tawny frogmouths have only one carotid located on the left side of the neck, they cannot turn their heads as far as owls (with two carotids) or they would interrupt blood supply to the brain. It would therefore seem possible that owls may be better at locating a sound source than tawny frogmouths and frogmouths may therefore need to use vision and hearing in combination. This is possibly the reason why tawny frogmouths, more so than owls, seem to forage most actively in the twilight hours.

Smell, taste and tactile senses

Compared to auditory and visual communication, relatively little is known of the importance of the sense of smell (or olfaction) in birds.[21] One of the problems for researchers is that smell, taste and tactile senses are far less easy to observe compared to vision and audition and any conclusive data on this sense have to come from detailed experimentation in conjunction with anatomical and physiological studies that can establish what receptor cells are available and which ones can be activated.[22]

Sense of smell and the nares

The nasal cavity has three important functions. Olfaction is just one of them. Another function is filtration of airborne particles, and a third function concerns thermoregulation. There have been studies to confirm that the sense of smell is important for foraging in some shearwaters and petrels[23] and odour discrimination has also been shown in domestic fowl[24] tested for different odour groups[25] and in some songbirds.[26] Mammals and reptiles may use their sense of smell for one, two or all of three main reasons: to find a mate, to find food and to identify marked territory. Usually, however, there is some trade-off in nocturnal species between the senses and especially between hearing, vision and smell.

It is not known to what extent a sense of smell is developed in tawny frogmouths. Interestingly, tawny frogmouths are probably singular in the degree to which their faeces smell. Using faecal expulsion to ward off predation (see Chapter 5) may be an excellent way of confusing reptilian predators relying on smell but this does not mean, of course, that tawny frogmouths are aware of the olfactory properties of their own excretions, although this is not impossible. However, their nests are never soiled and their faeces are always deposited over the edge of branches or over the edge of the nest, so there is little opportunity for tawny frogmouths to get acquainted with the odour related to their own faeces.

The nostrils do not just pick up scents but they are important for thermoregulation. The latter, of course, is a separate function unrelated to a sense of smell, but here may be as good a place as any to point out an important and specific feature in tawny frogmouth nestlings.

Conserving water is crucially important when hatchlings develop and the nasal cavity plays a very important role in water and heat economy in the body. King and McLelland[14] point out that recovery of water via the nasal cavity can be over 70 per cent at room temperature and 50 per cent when the ambient temperature is 30°C. In fact, for hatchlings in general, including tawny frogmouth hatchlings, water conservation is vital because they do not obtain any fluids other than via the food delivery by parent birds (that is, water is derived mainly from free water in food items but also metabolically derived from oxidation of foodstuffs).

The new hatchlings are equipped with two protrusions from the beak, both of which disappear over time. One is the egg tooth (see next chapter for more detail) and the other one is a modified keratinised flap (operculum) that largely covers the opening of the nostrils. It is quite an elaborate dome structure, with a downward opening, positioned neatly on top of the nostril (Fig. 3.10). It may be speculated

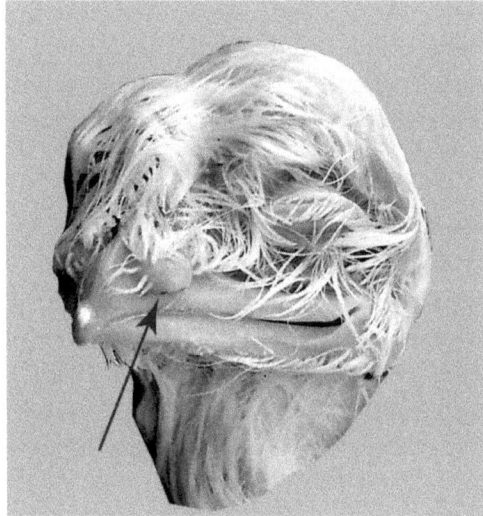

Fig. 3.10. Nasal cap (as marked by arrow) over the nostril in a hatchling. The dome structure is open at the bottom end and this might well assist in aiding water retention (minimising evaporation). The egg tooth (white on the upper mandible) is also clearly visible.

that this not only prevents rapid evaporation of water vapour, but may actually increase water condensation. The process of breathing in conjunction with water acts like a reverse-cycle air conditioner and aids in the cooling of air. Over the nestling and fledging period, this modified flap shrivels markedly (see Fig. 3.10 above showing an older bird) and by the time the juveniles are nearing adulthood this operculum has completely disappeared.

Although the shape, size and location of nostrils differ markedly between birds, there are likely to be only very few species that have any such obvious protrusions as part of the respiratory and olfactory system. Some shearwaters and petrels have tubular structures covering their nostrils and these are designed to filter out salt. Most non-Australian frogmouths are tropical birds and live in areas of ongoing high levels of humidity and it would be difficult to think of reasons why extra protection of the nares might be needed. However, the Australian tawny frogmouth inhabits some of the driest regions on Earth. It may well have evolved this nasal protrusion as an adaptation to arid conditions. As far as I know, this unique structure has never been described and there is no reference to this made with respect to other frogmouth species. It is not clear to what extent this dome-shaped structure covering the nostrils has to do with olfaction. Thermoregulation will be discussed generally in the next chapter.

Taste and touch
The senses of taste and touch have not received the same attention in birds as in other vertebrates although they had been described by a French researcher, E.

Goujon, in the 19th century.[27] He noted that parrots had receptors around the bill that did more than just provide some limited tactile information. These receptors also provided information on taste and were so densely packed on the inside of the upper and lower mandible (more densely placed on the lower part of the beak) that he referred to the structure as the 'bill tip organ'. Gottschaldt[28] noted that the bill tip organ is widespread among bird species that use the beak for selecting and manipulating food. One of the important functions of taste buds and the bill tip organ is to assess the palatability of food and to warn the individual of potential risks in consuming a specific food item. Tawny frogmouths consume a variety of poisonous invertebrates that are obviously not dangerous or unpalatable to them but one wonders if their tolerance to such invertebrates is not also helped by a reduction in taste sensitivity.

It is not known whether tawny frogmouths have developed a well-defined sense of taste. The available area for taste buds on the tongue, where most taste buds are found (especially at the base of the tongue and the tongue flaps) is likely to be of no use to tawny frogmouths. The tongue is a paper-thin triangular structure that is extremely small in comparison to the size of the beak (Fig. 3.11).

Moreover, not just the tip of the tongue and epithelium are cornified (as in many other avian species) but so is most of the surface of the tawny frogmouth's tongue. Such surfaces usually cannot accommodate receptor cells. In other birds, there are usually taste receptors in the lower jaw (at the floor of the buccal cavity) in close association with anterior salivary glands, as well as in the palate. It is possible that taste receptors might exist in those locations in the tawny frogmouth but this has not yet been studied.

Fig. 3.11. The tongue of a tawny frogmouth is paper-thin and triangular shaped. It does not appear to grow much. Here, pictured in a nestling, it takes up the length of the lower mandible but in an adult it is half the length of the lower mandible.

Apart from the bill tip organ there are also mechanical tactile receptors, called 'Herbst corpuscles' (sometimes spelled 'Herpst'). These are found at many sites on the body surface of most species. In some avian species these receptors are concentrated at the bill tip rather than along its edge, suggesting that they assist in actual prey detection. Tactile cues may also be gained via the group of long rictal bristles around the margin of the upper mandible. This is known to be the case in Australian ravens, nightjars, and all frogmouths and relatives, including the tawny frogmouth.[13]

Tactile information may be important in avoiding danger or to identify food. In social bonding and communication among some birds, tactile contact may be very important. Pair-bonded tawny frogmouths, for instance, engage in preening and this evidence of allo-preening would suggest some development of a tactile sense. Since the male preens the female, and does so only around the head region, it is less likely that this preening serves any form of ecto-parasite control as is the case in primates.

There is obviously a good deal more to be learned about the perceptual capabilities of the tawny frogmouth. Apart from the visual system, no other senses have ever been investigated. Most intriguing of all is their ability to move during the day and night with equal agility and responsiveness, as the next chapters will show, suggesting that their specialised nocturnal adaptations have not diminished their daytime mobility. These birds are equipped with a perceptual apparatus fit for all occasions.

4

Daily life and adaptations

Tawny frogmouth behaviour still poses many questions. This and the following chapters are by no means exhaustive but they provide a window into the night and day life of tawny frogmouths. Much is intriguing and unusual about tawny frogmouths.

Lifespan and interventions

It is not known how long tawny frogmouths actually live in the wild. Banding records of a small sample of 107 recovered birds showed that the oldest was 165 months or 13.75 years of age.[1] Some zoo records might have information on lifespan in captivity but this raises further questions. Captivity can have seriously detrimental effects on lifespan (shortening it) or, at times, may prolong life well beyond normal life expectancy in the wild, making the two sets of data not easily comparable. The data on captive tawny frogmouths, as sparse as they are, certainly suggest a potential for a very long lifespan, if one thinks of the recent report from London Zoo of having to euthanase a tawny frogmouth at the age of 32 years. And then there is the individual adult male rescued in 1994 in New South Wales that is still alive in 2017 at the time of writing the revised copy of this book.

Getting accurate records of optimal lifespan in the wild is not easy for any species. Many birds die well before they reach reproductive age and even those who survive on their own into adulthood can meet with misadventure, especially in modern society with its many civilisation risks. Tawny frogmouths feeding on roads at night regularly get hit by cars. There are many factors that can foreshorten the life of a bird. However, it is worth noting that Australian birds in general tend to have far longer lifespans than birds in the Northern Hemisphere. For instance, magpies may live for 25–30 years, and many other large Australian birds have been known to live to phenomenal ages of 80 years or more (galah) and even to 100 years of age (sulphur-crested cockatoo), at least in captivity or as pets. Bearing this in mind it is conceivable that the tawny frogmouth's actual optimal lifespan may be much longer than current records suggest.

In the records of tawny frogmouths, there appear to be some contradictions based on scant reports. Records apparently show that, at most, 30 per cent of tawny frogmouth offspring make it to adulthood and survive the first year.[1] More on this later. Suffice it to say at this point, that the overall figure of 70 per cent perishing, often before fledging, would seem remarkably high, although not entirely uncommon among avian species. It was established some decades ago[2] that only 14 per cent of magpies make it to breeding successfully, a sign that even for the best equipped native birds with abundant and amazing physical adaptations and resourcefulness, those that manage to raise offspring successfully are in a minority.

Predation and survival

There are many serious obstacles to overcome in order to survive. First, there are many other birds (among them ravens, butcherbirds and currawongs) that will attempt to steal eggs or kill young birds, mostly to feed their own young, which have a high need for protein during the nestling stage. Currawongs switch largely to fruit as adults and ravens are omnivorous, but for the intense growing bouts of nestlings, high protein food is the preferred food that parents will provide. Birds of prey, such as goshawks, hobbies and falcons, also take eggs and nestlings. Owls are as much a risk to adult tawny frogmouths as to nestlings, even though the boobook owl is slightly smaller than the tawny frogmouth. Rodents, particularly bush rats, collect eggs and regularly cause major damage to clutches of small and large birds alike.

Further, tree-climbing snakes can take eggs and nestlings. Carpet snakes (pythons) in particular can find and take even adult tawny frogmouths during the day.[3] On the coast, where some observations were made in areas with a good population of lace monitors (Fig. 4.1), it is rare for even larger birds to raise any of their clutches in any given year. In the hinterland of Woolgoolga in patches of remaining rainforest, I collected data on the reproductive success of kookaburras, currawongs, magpies and tawny frogmouths for more than 10 years. In the period of observation, the magpie pair managed to raise three offspring (one in 2001 and another two in 2013) and the kookaburras two offspring (2005). The currawong pair has never yet raised a single chick of its own species, although the pair once raised a chick that turned out to be a channel-billed cuckoo, able to vocalise in ways very similar to begging calls of currawong nestlings. A single tawny frogmouth appears to live a solitary life there and has never even been seen with a partner. Pied butcherbirds were more successful, raising one offspring successfully about every two years. Magpies and currawongs, in particular, get heavily parasitised by channel-billed cuckoos and the common koel. Moreover, the clutch is often taken at egg stage and nestlings are also not safe from lace monitors. Lace monitors are superb tree climbers and very swift in capturing prey (Figs 4.1 and 4.2). Sometimes tawny frogmouths, and other bird species in more subtropical

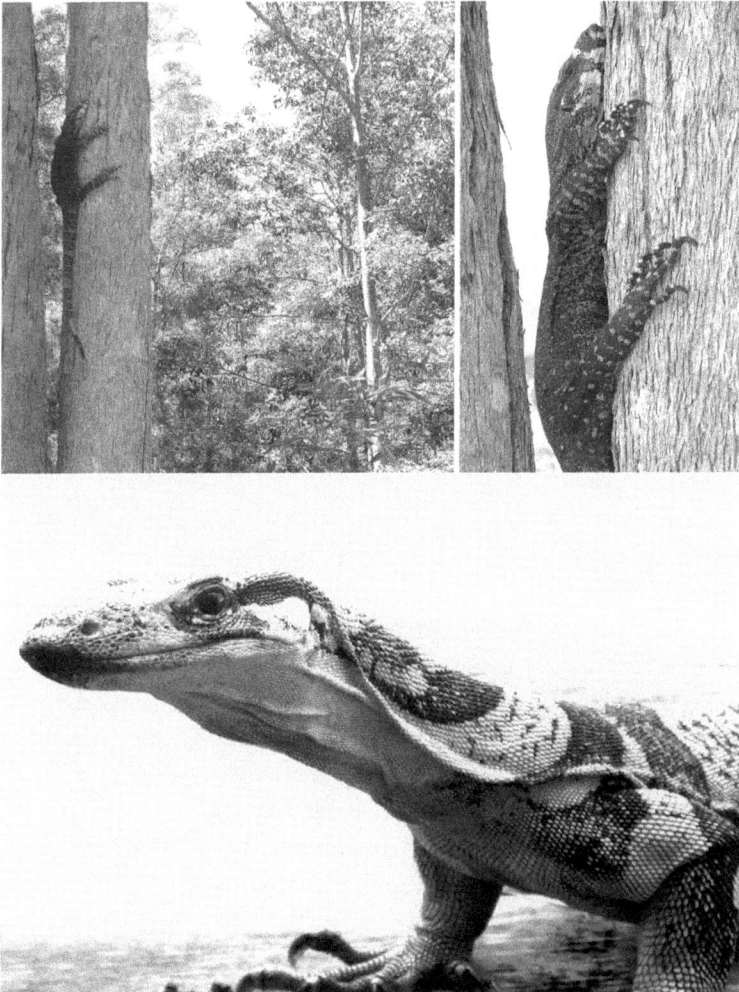

Fig. 4.1. The lace monitor is an extremely alert and capable hunter and a superb climber of the tallest trees. It can even jump short distances from one tree trunk to another and therefore has no difficulties at all in reaching any nest site close to the stem. This includes those of nest hole dwellers such as kookaburras, rosellas and other parrots and cockatoos (and also arboreal marsupials such as gliders) and tree stem nesters such as the tawny frogmouth.

Fig. 4.2. A lace monitor devouring a magpie juvenile from the feet up. Tawny frogmouths are also a prey item for lace monitors, not only at egg stage or early nestling age but even almost to adult age. Lace monitors also feed on the nests of possums and gliders.

areas where food is plentiful any time of year, may start brooding early in winter in order to miss the re-awakening of lace monitors hungry after bouts of hibernation. This is only a factor in southern latitudes and not in the northern tropical regions where lace monitors remain active for most of the year.

Lace monitors will feed on invertebrates and any of the smaller range of vertebrates (up to the size of ring-tailed possums), dead or alive. Nestlings are particularly easy targets. I have seen a juvenile tawny frogmouth being eaten and have also seen a juvenile magpie being devoured by a lace monitor but here I was far too late on the scene. Half the body was already ingested while the terrified magpie cried like a baby, a sound that is difficult to forget (Fig. 4.2). By contrast, the juvenile tawny frogmouth was silent while its wing was being eaten and only the half-open beak, the wide-open eyes and the raised feathers on the head indicated that the bird was very much alive. As will be discussed in the last chapter, tawny frogmouths have, in fact, very loud distress, fear and warning screams but, surprisingly, in this instance, the bird made no sound whatsoever.

Apart from natural predators and poor weather conditions, the most serious problems for tawny frogmouths arise as a result of contact with human society and its pets. Cats are the chief villains, according to some reports, but dogs and foxes also occasionally kill tawny frogmouths.[4] Many tawny frogmouths are killed on rural roads,[5] and an unknown number are poisoned by ingesting pesticides.[6] Sometimes this is indirect and death is substantially delayed because some toxins (administered to cockroaches or rodents) are absorbed into fatty tissue and the birds survive without any overt signs of ill health. Then, as the lean winter months wear on and as the birds start drawing on their fat deposits and lose weight, the poison that was previously dormant enters the bloodstream and the birds suddenly die. Winter months can be carnage, with spates of dead adult tawny frogmouths being found as a result of toxins ingested earlier in the year in a warmer season. In other words, life is made rather difficult by the addition of countless hazards and dangers in addition to natural predators, diseases and poisons contained in the environment (see Chapter 5). However, in some pockets, there seem to be particularly favourable conditions in which food availability and nesting opportunities are good and survival to fledging state has been shown to be high.[7]

Roosting and camouflage

Roosting during the day and camouflage go together in tawny frogmouths. However, arguing, as Fleay did,[8] that they never seek shelter and, more often than not, roost on bare branches, is a claim not confirmed by my observations and by other studies.[9] Indeed, it was not uncommon for tawny frogmouths to change day roosts when the sun became too hot or a sheltered area too cold. Boobook owls always seek shelter under well-leafed trees to avoid being seen and mobbed by diurnal passerines. Tawny frogmouths appear similar in size and head shape to boobook owls and can therefore also expect to be mobbed unless they adopt a camouflage posture (explained below). Butcherbirds and noisy miners seem to have a particular dislike of tawny frogmouths, especially juveniles, and will mob them with great regularity.

Here it is necessary to distinguish the behaviour of tawny frogmouths on day and night roosts. Day roosts, of whatever kind, are associated with camouflage posture, whereas night roosts are usually not. I have never once seen tawny frogmouths in camouflage posture at night, and that is over a 20-year period. Hence, it is relatively safe to deduce that camouflage postures are designed to protect tawny frogmouths during the day.

Many writers have commented on how well the markings and colouration of tawny frogmouths work to conceal them (Fig. 4.3). Their facial strobes around the eye further make the bird look more like a branch than a bird.[3] In order to work as camouflage, however, the bird needs to enlist the help of its environment. It needs to have in its immediate vicinity trees that best match its colouration and markings and roost on a branch of such a tree or near a trunk, otherwise the concealment will not be effective (Fig. 4.4). A systematic study[9] confirms this, showing that

Fig. 4.3. Nestlings in their protective camouflage plumage.

tawny frogmouths exhibit a significant roosting preference for the coarse and dark-barked stringy-bark trees but will also frequent smooth-barked, light-coloured gums. If the birds choose the latter, they will then preferentially sit on dead branches usually having a coarser structure and, generally, darker colour than living branches (Fig. 4.5).

Adopting a camouflage posture consists of several movements. It involves stretching the entire body and the head upwards (the beak facing upwards giving the appearance of a branch that is petering out), sleeking down all feathers on the head and body, and closing the eyes to a small, inconspicuous slit (see Fig. 4.7). Once that is achieved, the next phase of the camouflage posture is to stay absolutely still and remain so until the perceived danger is over. There is one added requirement. In many cases, the camouflage posture also has to be angled in the way the tree branches are growing and here the function of having a semi-zygodactyl second toe (as described in Chapter 2) becomes abundantly clear. In order to achieve maximum benefit from the cryptic colouration of its plumage, the bird has to align itself in the direction of the branch, not at a right angle to it (as normal perching would be) and for this it is important to have mobile digits that can fit around the branch lengthwise. If a pair of tawny frogmouths roost together, they do not align themselves in the same direction but will branch out in opposite

Fig. 4.4. The familiar and often described 'stick' or camouflage posture of the tawny frogmouth. The angle of stretching can vary from almost vertically upwards to an angle of nearly 45°. This angle appears to be dependent on the direction of tree branches of the roost.

directions as a mirror image of each other. The effect is that the two birds together now look like the fork of a tree with two branches going in opposite directions but at about the same angle.

The clear preference for day roosts that provide a colouration and patterning similar to their own suggests that the theory of roosting as 'camouflage' is correct. Moreover, the actual physical change to a camouflage posture seems to be genetically embedded and a reflex response. If an unknown stimulus is presented, even two-week old nestlings will engage in the camouflage posture; not very successfully at first, however, because their extensive feathers on the head do not permit the same sleek look as is typical in adults (Fig. 4.6). Choosing the right tree for roosting, by contrast, might well be a learned behaviour because fledglings tend to choose entirely inappropriate roosting spots, at least for the first four weeks post-fledging, and will land and stay on surfaces that expose rather than hide them.

Fig. 4.5. A beautiful demonstration of choice of substrate for a day roost and the protection that the tree affords. The bird is stretching slightly in a camouflage posture and the feet are angled up the trunk.

But is it correct to say that the camouflage posture helps to avoid detection by predators? The camouflage posture is triggered only when a tawny frogmouth becomes aware of an approaching or moving object. That may be a human or a bird of prey. When the tawny frogmouth begins to change its posture, it may well be too late to avoid detection, particularly since changing from a relaxed sitting and sleeping position to a camouflage position involves movement – something predators are very quick to detect. It seems, therefore, rather more likely that the roosting position itself (blending in with the environment) is more important than camouflage posture changes in avoiding detection by an aerial predator. This is possibly different for ground predators where camouflage posture changes might assist tawny frogmouths in avoiding detection. The camouflage does not work at all in the case of reptilian predators because they rely first and foremost on smell. Avoidance of reptilian predators is easy, however, and such predators only constitute a risk to a seriously injured adult frogmouth. Indeed, tawny frogmouths, like kookaburras, can actually kill snakes but, unlike kookaburras, will generally not eat them, unless they are very small.

Tawny frogmouths change day-roost positions surprisingly frequently. Although they will usually remain on the branch of the tree that they occupy at daybreak, they

Fig. 4.6. A nestling will already adopt the camouflage posture when approached. This one, pictured in front of its parent, had its eye nearly shut. Source: Chris Watson/Shutterstock.

change day-roosting positions from one day to the next. It is relatively rare to see them roost in the same place for more than three days running.[9]

Day roosts and thermoregulation

During day roosting, harm may come to individuals in two ways: from predators and thermoregulatory cost. Sitting all day out in the open, particularly when sitting on bare branches, raises the question of protection from high temperatures in summer months and from cold temperatures in winter months. It would, therefore, make sense for tawny frogmouths to choose positions that do not have all-day exposure to the sun when the sunrays are at maximum strength and supplement warmth by more exposure to the sun when it is cold. Although tawny frogmouths do not always follow this rule, research results have found a significant difference between northerly orientation in winter (more exposure to sun) and protected areas from the sun in summer months.[9]

During daylight hours, tawny frogmouths can sometimes be seen on the ground and passers-by make the mistake of thinking that the bird is injured because it is splayed out, motionless, with wings wide open, beak wide open and eyes shut. But this is not so. Tawny frogmouths go to the ground, even when not feeding, in order to take a sunbath. The warmth of the sun has a strangely paralysing or mesmerising effect on birds (and not just on tawny frogmouths but magpies, currawongs, indeed many avian species). The bird also moves the head to the side so that the sunrays can penetrate beneath the thick layer of feathers, and may remain motionless for up to five minutes at a time.

This sunbathing behaviour may play a role in thermoregulation, but it may also be related to two other possible benefits. One is the direct administration to the skin of sunrays that will then allow the body to synthesise vitamin D. Another possible benefit is parasite control, either by exposing larvae to more sun than they can sustain or by inviting pests to leave the body. Parasite control is thought to be achieved by dust bathing but, to my knowledge, tawny frogmouths do not use dust bathing.

When there is no sunshine, the reason for coming to the ground during the day[10] is difficult to determine. Why they would do so at times when not feeding is not clear. On those occasions, they often have their wings outstretched as if to support their bodyweight but their eyes are open. I have observed this behaviour on several occasions. Eventually, they will fly up to another roost. It is possible that tawny frogmouths once regularly roosted on the ground, just as their cousins the owlet-nightjars and the nightjars sometimes still do, and were forced back into the trees for roosting because of the multitude of introduced mammalian predators (cats, dogs and foxes). The Eastwood State Forest, in which University of New England researchers conducted their study, is infested with all three ground predators and neither researcher had ever seen a tawny frogmouth roost on the ground.[9]

Sleeping

How is a nocturnal bird able to sleep and rest during the day without getting killed? Actually, having observed them for long periods of time during the day it seems that tawny frogmouths, and other nocturnal birds, rarely ever completely sleep. In fact, all animals, particularly prey animals, suffer all their lives from a fundamental conflict between sleep and wakefulness because sleeping is a dangerous activity. It appears from recent literature that only birds (and possibly some reptiles) have solved this conflict by engaging in unihemispheric sleep. This is a unique state in which one brain hemisphere sleeps while the other remains awake. Rattenborg and colleagues[11] have investigated behavioural, neurophysiological and evolutionary aspects of unihemispheric sleep in birds and have shown that unihemispheric sleep may be chiefly concerned with predator detection. They observed mallards (ducks) sitting in a row and found that any duck in the more vulnerable position at the end of the row (not protected from conspecifics at least on one side) showed an increase of 150 per cent in unihemispheric slow-wave sleep and responded rapidly to visual stimuli presented to the open eye.[12]

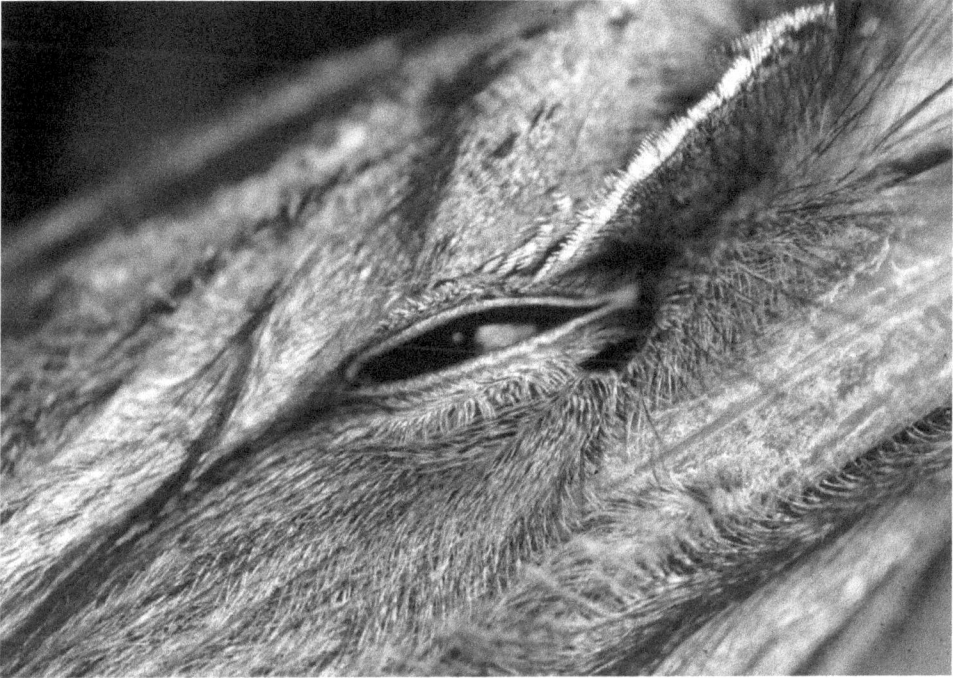

Fig. 4.7. Tawny frogmouths can keep their eyes open a fraction and still be asleep. This allows them to respond to movements. The detection of movement results immediately in the adoption of the camouflage posture.

Potoos, distant relatives of the tawny frogmouth, and very similar in appearance (as stated in Chapter 1), have solved the problem by the morphological adaptation of having a vertical slit in the eyelids so that they can see even if the eyelids are closed completely. Tawny frogmouths do not have this adaptation but they tend to keep their eyelids open just a fraction even when seemingly asleep (Fig. 4.7). This ability to sleep without closing the eyelids completely appears to be a physiological adaptation because even two-week old nestlings do this quite effectively (see Fig. 4.6). It is very nearly impossible to approach a resting tawny frogmouth without triggering a camouflage response.

The longest sleep periods I have observed occur in the early morning hours, just after hunting and feeding and, oddly, an hour or so at dusk just before tawny frogmouths become extremely active. Not all writers have made these observations but, at the very least, it can be argued that there may be a great deal of variability in nightly foraging habits (as discussed in the next chapter).

A state of wakefulness is easy to observe. I have watched tawny frogmouths roost for many hours and discovered that they actually sleep surprisingly little during the day. Short bursts of sleep are followed by hours of preening, looking around and, occasionally, even by flight to another roost. The number of hours spent awake are more than the hours spent sleeping. Total sleep during the day rarely exceeded more than four hours. However, I have also observed that they actually sleep at night and these sleep sessions seemed to fall into short bursts within a two-hour period before midnight and a two-hour period in the very early hours in the morning, between 2 am and 5 am. The patterns of sleeping briefly during the day and night have been observed in juvenile, hand-raised tawny frogmouths as well as in free-roaming adult birds. At best, then, the tawny frogmouth is a sporadic sleeper cycling between states of sleep and wakefulness over a 24-hour period.

Young and recently fledged tawny frogmouths do many things that keep their parents highly stressed, such as fly and flutter repeatedly during the day, sit on the exposed roofs of houses, and even vocalise, making themselves dangerously conspicuous. All the parents can do is follow them, sometimes trying to edge them back into a more sheltered position. Both parents may flank the young tawny frogmouth and eventually get it to settle down. All this commotion may happen in broad daylight, again suggesting that the tawny frogmouth has eyesight that works well during the day.

Another one of the antics of a young tawny frogmouth is to feed during the day and even in broad daylight. I have seen several recently fledged tawny frogmouths taking to the air and snatching flying beetles near eucalypts in swift and almost vertical upward flight. The day feeding and awake states have unusually been associated with neurological disease but the juvenile tawny frogmouths that I had opportunity to observe regularly for over a month post-fledging were healthy and well.

Coping with the weather and the seasons

Tawny frogmouths must rank among the most exceptional all-weather bush birds. Their feather insulation is so complete and perfect that neither cold nor heat, nor even rain and storms, appear to affect them greatly. I have watched tawny frogmouths on their day roost in summer at temperatures of over 30°C and have seen them remain motionless without any signs of physiological discomfort for continuous periods of exposure to the sun of eight hours or more. When the temperature rose to well over 30°C, other bird species around them showed panting behaviour (half-open beak) and, quite often, they spread out their wings in a tent-like fashion. Both strategies are designed to cool a bird. The tawny frogmouths, however, did not even open their beaks for better ventilation or seek shade and gave no indication that they needed to cool down. Physiological tests conducted in response to heat stress have shown that tawny frogmouths can triple their breathing rate without needing to open their beaks and only when their actual body temperature rises by as much as 4–5°C will they pant. At any further heat stress, they can engorge the blood vessels in the mouth area and produce a mucus that, in turn, helps to cool the air as it is inhaled and thus cool the body. But these were experimentally induced conditions.[13] Still, importantly, there has been a spate of papers on basic metabolic rates in the frogmouth family (Podargidae)[14] confirming that frogmouths are the avian family with the lowest metabolic rate (energy expenditure at rest).[15]

Coping with heat is one thing but what about the cold? Do the feathers suffice to protect a bird from the extremes of cold experienced in many regions of Australia at night, including deserts? And how does the bird stay and survive when winter months impose freezing temperatures? On the New England Tableland (~1000 m altitude), where most of my long-term observations took place, temperature fluctuations of up to 25°C between day and night are common. In winter months, night temperatures can plummet to as low as –13°C. There have been extreme nights with temperature drops down to –18°C and I found tawny frogmouths in the morning alive but with small icicles hanging from their eyebrows, beak and bristles. How can an animal exist in such an environment and why doesn't the tawny frogmouth have to migrate during winter?

Small birds, such as hummingbirds and small birds in arctic and subarctic regions,[16] go into torpor but tawny frogmouths seemed too large to do so, until it was confirmed that tawny frogmouths also, and regularly, use torpor as a physiological mechanism to save energy. Their body temperature can fall to as low as 29°C if only for relatively short periods of time. Dawn torpor bouts are shorter than night torpor bouts and temperature reduction may be as little as 0.5–1.5°C. However, night torpors could be long and extensive and the greatest reduction of body temperature was ~10°C, down to 29°C; shorter reductions usually resulted in the lowering of the body temperature by 5°C.[17] The problem is that tawny frogmouths often cannot get very nutritious food, especially in the winter months, hence energy conservation is of paramount importance. Torpor is not just a regulatory mechanism to cope with loss of heat, but assists in the energy regulation of an organism when food intake is low.[18] It was found that torpor frequency, depth and duration were greatest when arthropods were less abundant, establishing a link between food availability and heterothermic responses.[19]

There may be differences between frogmouth species generally, since most frogmouths are tropical species, and variations in heterothermy may also be dependent on roosting sites, food availability and other factors.[20] Whatever these factors may be, this ability to move the body in and out of torpor must be one of the most astounding adaptations in a large bird.

5

Feeding and territory

Key elements used to distinguish nocturnal from diurnal species are the timings of their active foraging, travel and socialising, but close examination of tawny frogmouth behaviour somehow blurs the line between diurnal and nocturnal species. As already shown, tawny frogmouths may move about and be awake during day and night. Feeding, however, is largely a matter of dimming light or night, with a few observed exceptions when beetle swarms invite some hawking flights even during daylight. Generally, when daylight is fading, tawny frogmouths start changing roosts and by the time twilight has descended, they are actively beginning to look for food.

Ultimately, the extent of their foraging activities depends on the quality and type of territory. Quite extensive dietary records are now available, obtained from examining the stomach contents of dead birds. However, there has been some disagreement regarding how, and even when, tawny frogmouths hunt. In some accounts, tawny frogmouths are referred to as 'crepuscular' feeders, simply meaning 'twilight' feeders. Other records[1] claim that tawny frogmouths hunt through the night.

The question of when they feed ought to remain relatively open. Clearly, as was shown in the last chapter, roosting and sleeping alternate and, after dusk and before dawn, sleep alternates with foraging expeditions. When raising chicks in the nest, the parents start feeding at dusk. They usually take their first break after two hours and then continue feeding well into the night. In one set of observations, 17 feeding visits were recorded in a span of four hours.[2] Raising nestlings and feeding fledglings will be further discussed in Chapter 8. In general, however, tawny

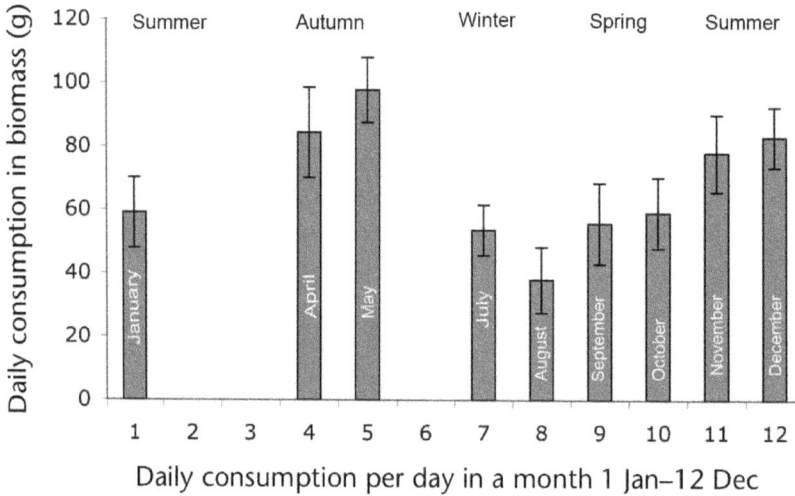

Fig. 5.1. Consumption of food provided to one free-ranging tawny frogmouth. Note the sharp increase in food intake in the autumn in preparation for winter. Interestingly, the food intake was reduced in winter. This reduction in food intake in July and August (significantly lower in winter than in summer and autumn) was, in this case, not contingent on food availability. The offered food was mostly rejected. Hence, they may eat less not because of food scarcity but because they have physiologically adapted to winter food shortages. The missing data points (February, March, June) correspond to times when the tawny frogmouth could not be located.

frogmouth chicks are not different from diurnal chicks. It depends on the biomass that is being fed but, as a rough guide, offspring are fed every 20 to 30 minutes in diurnal birds and the same is true of nocturnal birds. There are breaks built into this and while a break in feeding tends to be just after midday in diurnal birds, it is around midnight in nocturnal birds. Foraging excursions and feeding times also vary with biomass obtained, energy expended and ambient temperature.

To obtain at least some idea of the amount of food consumed by a tawny frogmouth, I collected data using birds prepared for release. While the measurements may not tell the whole story of food consumption, they do show trends and, moreover, the biomass indicated in Fig. 5.1 corresponds very closely to the food intake of a male that was not deemed fit for release.

Figure 5.1 shows a substantial increase in the daily consumption rate in May (before the onset of winter) and significantly lower consumption rates in the middle of winter (July and August). The amount of food consumed in autumn was staggering, translating to a biomass of about eight adult house mice per 24 hours. Presumably, autumn is a time to build up fat reserves in preparation for the time of food shortage in winter.[3]

Type of food consumed

The food of tawny frogmouths, despite comprising enormous variety, can be simply explained: it consists mainly of things we regard as vermin in houses and as pests on farms and in gardens. In fact, tawny frogmouths may be among Australia's most effective and useful pest control birds. In order to make up sufficient biomass, large numbers of invertebrates are consumed. These include a variety of insects such as bugs, cockroaches, beetles and moths, as well as spiders, centipedes, millipedes, scorpions, caterpillars, snails and slugs. Importantly, they can feed on terrestrial *and* aerial invertebrates.[4] Among the vertebrates known to be eaten are frogs and mice.[5]

This list of items is largely derived from two sources: from a book describing the diets of all native birds[6] and from the examination of stomach contents of 40 tawny frogmouths.[7] The researchers' findings can be summarised by saying that the majority of a tawny frogmouth's food consists of insects (78%); spiders and centipedes makes up a further 18 per cent and the remainder (4%) are vertebrates, such as frogs, lizards and mice – all terrestrial species.

Some particular food items – snails, slugs and Christmas beetles – deserve more explanation. These food sources are quite important but for entirely different reasons. Slugs and snails may not leave any trace in the digestive system, but, as will be shown later, are quite important (and sometimes fatally so) for the tawny frogmouth. Slugs (see under the heading 'Problems with food' later in this chapter) can lead to a deadly disease in tawny frogmouths.

Christmas beetles belong to the large genus *Anoplognathus* in the family Scarabaeidae. The adult beetles can defoliate trees and are a substantial pest in the Australian bush, having at times been hailed as the major cause of die-back. Christmas beetles feed on eucalypts and, apart from their hard shells, are indigestible or even poisonous to many other birds. While the tawny frogmouth is extremely susceptible to human-introduced chemical poisoning, it has a tolerance to the eucalypt-derived toxins that are consumed by Christmas beetles.

How such tolerance might have developed was not clear until I accidentally observed that nestlings and fledglings chew on eucalypt leaves. To establish whether this was merely some erratic behaviour or indiscriminate chewing, juveniles were presented with branches of eucalypts, upon which I had seen Christmas beetles feed, and with other leaf-carrying branches of introduced trees, such as Japanese maple. Without fail, the tawny frogmouths regularly chewed on the eucalypt leaves but left the maple leaves alone.

I have since seen this behaviour of chewing – not consuming – eucalypt leaves in at least 18 young tawny frogmouths. For a species that feeds on insectivorous and vertebrate species, this seems remarkable and unusual. Since this behaviour has occurred with great regularity, but exclusively in nestling or juvenile tawny frogmouths, one might perhaps speculate that by ingesting eucalypt leaf oils in small amounts at a certain stage of development, the young tawny frogmouths may develop the kind of tolerance needed for them to later feed on the toxic Christmas beetles. Without any notable ill-effect, young tawny frogmouths feed on Christmas beetles even during the day and late in the afternoon, snatching them in flight near the eucalypts around which the beetles hover. Still, it is hard to believe that a physiological advantage may be derived from chewing plants that may enable tawny frogmouths to feed on beetles.

Some writers have taken it upon themselves to describe the beak of the tawny frogmouth as a 'clumsy left-over of past aerial practices'.[8] This is a very strange comment indeed because aerial practice, as shall be explained further below, is not a past practice but one in which tawny frogmouths excel. Further, the beak is an instrument devised specifically to cope with wriggling, hard-shelled, stinging and biting invertebrates. The beak truly comes into its own when the tawny frogmouth feeds on invertebrates such as centipedes and even Christmas beetles. The massive beak, with its flat edges, pulps the hard shell of a Christmas beetle before swallowing. Catching and holding a prey item at the end of the beak, then pulping it on the side of the beak, the bird can avoid the toxins of centipedes, or even the sting of a scorpion. It is further worth noting that the bristles around the beak and the tuft on the top of the beak, described in Chapter 2, show their usefulness in a function that has never before been described. The bristles of the beak in nightjars were thought to serve the purpose of netting insects in flight. It is more likely that, in tawny frogmouths, the bristles are there for the bird's protection. I have observed that centipedes may try to bite but, if caught on the side of the beak, they get hopelessly caught in the bristles and tuft and can find no point of attack.

Water regulation

Tawny frogmouths are found in humid and arid zones of Australia. The problem for all native Australian species living on the driest continent is how to deal with water shortage, water conservation and evaporation. The third chapter has already dealt with most of the aspects of these pertaining to tawny frogmouths and the dome-shaped structure over the nostrils of hatchlings was described in detail, showing its effectiveness as a tool for water conservation in the body. There is a further twist to this remarkable biological phenomenon. In addition to their low metabolic rate, the ability to go into different states of torpor on demand and their efficient thermoregulation, one writer has observed that he has never seen a tawny frogmouth drink.[9] He was right; as large a bird as it is, the tawny frogmouth does not seem to drink. Most food items that tawny frogmouths consume also contain fluids, with the largest volumes, of course, in vertebrates. A food source rich in fluids apparently sustains most of, if not all, the bird's fluid requirements. There are many other insectivorous birds, such as magpies, that drink water at regular intervals, although perhaps not as often and possibly not as much as grain-feeding birds.

How and where they feed

There have been surprisingly categorical statements made about the way tawny frogmouths feed and where they forage. While most agree that they forage on the ground, one group has asserted that tawny frogmouths do not take food on the wing,[10] while another contends that they do.[11] Since I have witnessed tawny frogmouths taking insects on the wing with great regularity, it is not clear to me how this can be denied. Their aerial foraging flights are not hit-and-miss events. They consist of short snatching or hawking flights to foliage, branches or into the air.[7]

Tawny frogmouths use their beaks for catching prey with great precision. I can give a very personal example of this. At the time of releasing tawny frogmouths, I train them to take food from me in flight. To achieve an accurate outcome takes about an hour. A dead mouse is held at the end of thumb and index finger, and the hand is raised above the head for them to collect it in flight. These sessions barely qualify as 'training' because the birds usually just took two to three attempts at most to succeed in their aim-and-snatch flights, but this happened in relatively good light conditions. After they were released, any support feeding was only done after sunset and such support feeding usually ceased after four weeks post-release. The feeding was undertaken in a generally open area in which nearby tree roosts were in plentiful supply. With my hand (with mouse between the thumb and index finger) raised high above my head, I never had to wait for more than 10 minutes, mostly far less, before the mouse was snatched from my hand. Impressively, while the mouse only cleared my thumb by ~10–30 mm, the beak of a tawny frogmouth never touched my thumb. My hand was never knocked and the beak never once scraped my skin. The rate of accuracy was astounding. Moreover, it was often pitch dark. I could not hear nor see where the tawny frogmouth was or from which angle it might approach. The only evidence of the presence of a tawny frogmouth was indeed the fact that the mouse was taken in one swift move and with the lightest possible suggestion of displaced air from a wing beat near my head. Even under extremely dim light conditions, in which the human eye can barely distinguish the outline of trees, the tawny frogmouths were able to collect the food with unfailing accuracy.

Tawny frogmouths are sometimes thought of as clumsy. However, collecting food in flight with such unfailing accuracy is anything but clumsy and shows a very accomplished hunter. In hindsight, of course, it is not surprising for a hunter to possess such skills. What is surprising is that, at times, it has been doubted that tawny frogmouths hunt on the wing.

In general, it is true that tawny frogmouths are sit-and-wait predators,[7] as are kingfishers (including kookaburras) and some owl species. However, when they go to the ground for feeding, I have seen many examples of such landings that were not motivated by having seen a prey item. Once on the ground, they may search a small area before they take a short flight to another patch. As frogmouths feed on some ants, cockroaches and ground spiders, they may be assured of finding something, but such small prey items are not usefully obtained by a sit-and-wait method. In some more heavily forested areas and wetter regions, it has also been found that tawny frogmouths feed extensively on frogs. The naturalist Edith Coleman watched tawny frogmouths pouncing on frogs, presumably to stop the frog from hopping away, but saw tawny frogmouths hawking for insects only once or twice.[12]

It has also been observed that tawny frogmouths may move to the ground after a wildfire. After such a wildfire on Weddin Mountain in New South Wales, two tawny frogmouths were observed feeding on the ground in burnt areas.[13] Presumably, such fires may leave lizards and larger insects behind that have not completely burnt. This observation also discounts claims that tawny frogmouths only feed on moving objects.[10] Neither the mice that have been offered as food nor many of these insects are moving, yet they are obviously recognised as food. Moreover, cicadas, grasshoppers and spiders are often collected from trees and, more often than not, if the insects are diurnal species, they do not move at night.

Ground feeding has also been observed in towns, ports[14] and on football grounds where tawny frogmouths feed under floodlight. Such sites where the birds assemble could be larger than one territory. It appears that, if a rich and abundant source of food is found, strange or neighbouring tawny frogmouths may tolerate each other.

Interestingly, tawny frogmouths rarely approach prey they have spotted in a direct line of flight. The usual procedure is to get to one or even two more roosts closer to the location where a prey item has been spotted. When they do their last direct flight it is swift and accurate. Direct flights to collect food from my hand usually cover distances of 2–5 m.

Tawny frogmouths do not actually consume the item they have collected in flight or on the ground unless these are very small. The usual procedure is to capture the prey in the tip of the beak, hold it there and take it to the nearest convenient feeding branch. The food is then processed. This may involve pulping an insect at the rim of the beak before swallowing it or using the beak like a ramming tool and, with some force, bashing the catch against a branch in exactly the same manner as do kookaburras. Lizards and mice are usually killed in this fashion before being consumed.

I am not certain what Glen Ingram means when saying that tawny frogmouths 'gulp down' their food.[15] Most animals would do this to ensure its safe passage into the stomach. It is true that frogmouths swallow their food with great vigour. Presumably, it would be important for them to get food transported to the stomach as quickly as possible. This would make sense particularly in the case of stinging invertebrates such as scorpions and millipedes because stings can remain briefly active even after the organism has died. Tawny frogmouths have extraordinarily acidic stomachs.[16] It is probably for this reason that tawny frogmouths, unlike birds of prey, magpies, kookaburras and other avian species that regularly eliminate indigestible food in pellet form, only very rarely produce pellets.[9] For the tawny frogmouth, it seems, no food in its large repertoire of food items is too indigestible.

On the rare occasion that I have collected a large rounded pellet from the male tawny frogmouth, it was after several weeks of the bird consuming food containing high levels of roughage (bones, hair, moth wings) and in those rare cases the bird refused food until after the pellet had been expelled via the beak. All owls regularly produce pellets but they tend to be barely formed and usually consist of a packed assortment of hair, feathers and undigested bits of bone. By contrast, the tawny frogmouth pellets were smooth and about the size of a stewed plum. In currawongs, pellets expelled regularly have about the size and appearance of dark pips of olives, elliptic in shape. Their pellets have a further important function: since they are fruit eaters, they expel the seeds which are then provided with their own moist culture in which to grow. Currawongs are among the most important dispersers of native fruit seeds and it is the pellet that provides the vehicle for germination of the seeds contained in the pellet. In owls and tawny frogmouths, it is merely part of the digestive process. Presumably, it is an advantage to expel roughage via the beak than allow possibly sharp bits of bone to travel through the entire digestive system.

In summary, tawny frogmouths feed at all strata of the physical environment, in the air, in trees and on the ground and their techniques of capturing prey, such as picking food off the ground or from tree branches or taking their food in flight, show great versatility. Food processing methods, such as snapping, pulping, bashing and pouncing reveals that they adjust their feeding technique each time in accordance with prey type and prey availability.

Such versatility is remarkable and puts the tawny frogmouth in a class of its own. Many birds feed either exclusively on the ground or in the air, and quite a number of smaller songbird species are even confined to a specific layer in trees, being permitted to forage only in the canopy or the understorey (such as thornbills).

It is remarkable how often the location of food capture is not taken into account in biological control measures, and even in other contexts, despite it being so crucial to a species. For instance, the cane toad was introduced to Australia in an attempt to control the native grey-backed cane beetle (*Dermolepida albohirtum*) and Frenchi beetle (*Lepidiota frenchi*) that were devastating sugarcanes. The only problem was that the beetles occurred in the upper plant layer and the cane toad is a ground feeder. The same biological mistake was made when Australian magpies were exported to Taveuni, one of the Fijian islands, to control palm moths. Magpies are strictly ground feeders while the palm moths chew away palm fronds high up in the canopy.[17] Predator and prey items never met! Hence it is as important to understand the vertical space of feeding ranges as it is to understand the range of items they may consume. For birds as large as tawny frogmouths it is probably essential that they are able to access the vertical planes of a forest as much as the ground and the airspace.

Problems with food

Tawny frogmouths are also sometimes victims of pesticides and diseases that were introduced by rats. Tawny frogmouths are subject to great risks and die regularly from ingesting poisons, such as organochlorines. They are found moribund, weak, crying or convulsing during late winter and early spring of each year, especially along the east coast of Australia. The first report of possible organochlorine (OC) toxicity in a tawny frogmouth was published in 1981. A tawny frogmouth from Victoria was found to have elevated concentrations of lindane, heptachlor and α-benzeneheptachlor in the brain.[7]

A further investigation into tawny frogmouth mortality was undertaken in a collaborative effort between the Wildlife Information and Rescue Service Inc. (WIRES) and Veterinary Pathology Services. This investigation revealed elevated concentrations of four OC compounds within liver and brain samples collected from tawny frogmouths that had died suddenly in 1994.[18] In a 2005 report, 'Common Diseases of Urban Wildlife (Birds)', it was noted that these four chlorinated hydrocarbon compounds (oxychlordane, heptachlor-epoxide, DDE and dieldrin) have been used primarily as insecticides. Despite controls imposed on their use, some continue to be circulated in termite eradication while others, such as heptachlor, aldrin and dieldrin, are widespread in domestic cockroach and termite control.[19] Cockroach and termite control activities, so it was argued,[18] likely provided a persistent environmental source of OC for tawny frogmouths. There are also various rat and snail poisons that can be causes of death. The irony is, as other writers have also stated,[7] that tawny frogmouths feed on items that we want to be rid of, but instead of allowing this, by our intervention, it is tawny frogmouths that get killed. Tawny frogmouths can take prey types avoided by most other insectivores and this gives these birds a special niche among day and night birds. Tawny frogmouths are invaluable for general control of bush invertebrates, but it is that niche (and usefulness) that puts the tawny frogmouth increasingly at risk. By feeding on cockroaches, for instance, they suffer the same fate as their prey.[7]

Food intake has caused the tawny frogmouth another problem. Transmission of a disease called *Angiostrongyliasis* has increasingly made headlines in veterinarian literature.[20] It originates in rats, or more precisely, with the rat lungworm, *Angiostrongylus cantonensis*, a nematode parasite. It manifests as a neurological disease that has caught the attention of veterinarians and the medical profession alike because this infection has presented in humans, marsupials and birds. Histological sections of the brains and spinal cords of 13 of 22 tawny frogmouths from the northern suburbs of Sydney were shown to suffer from *Angiostrongylus cantonensis*. The tawny frogmouths were weak, unable to perch or fly and often unable to right themselves.[21] This parasite is now well recognised as the primary cause of eosinophilic meningoencephalitis in humans and is carried by black rats.[22] Rats eliminate hatched larvae in their faeces and the cycle continues via snails and slugs that forage on rat faeces, thus serving as intermediate hosts. Humans, tawny frogmouths and other animals can become infected through ingestion of this intermediate host: humans become infected by ingesting lettuce contaminated with slugs or their slime, and tawny frogmouths by eating the snails or slugs. That this disease is a serious issue affecting the long-term health and prognosis for survival of the tawny frogmouth, especially on the east coast of Australia, is now only too apparent, so much so that it has become a subject of reports at the international World Organisation for Animal Health in Paris and a good deal of further research, especially since its presence was detected in the Sydney area and in southern Queensland.[23]

The tawny frogmouth has been called 'a biosentinel'[21] and it may well be a modern-day canary in the coal mine, that is, a warning system to the environment in which we live. The tawny frogmouth is one of the species that has adapted to human habitation and has learned to make its territory in and around major city centres. It seems to be paying a very high price for that adaptation.

Territory and territorial defence

Tawny frogmouths have been described as 'remarkably sedentary'.[15] In one detailed study[24] territory size was more or less accurately determined over a period of several months. The seven pairs in this study group lived within a 219-ha area and the researchers thought that there might have been more tawny frogmouths within the same forest area. Size of territory was estimated at ~20–30 ha per pair. There have been suggestions that a single tawny frogmouth may need as little as 1.2 ha[25] but nowhere else in the literature has this claim been substantiated or confirmed by systematic study. Indeed, figures of around 20–80 ha per pair[26] appear to be more typical.

The size of a territory required to sustain a pair of large birds depends on many factors. There are at least three basic requirements. One is food supply. Ideally, a habitat needs to produce replenishable food at a rate higher than the tawny frogmouth consumes. Second, it needs a sufficient number of good roosting sites. Third, it needs sites that are suitable for nesting. The presence or absence of potential predators, especially some owls, might well also influence whether tawny frogmouths will stay or be forced to go elsewhere. Closeness to a water source may only be important as far as it concerns the presence or absence of classes of invertebrates and vertebrates, such as frogs and lizards, not necessarily as a source for water consumption.

Like many other territorial birds, tawny frogmouths stay in the same territory for as long as possible. The examples mentioned previously showed that a pair stayed in the same territory for six years. Upon the death of the male, the female stayed on and eventually formed a new partnership with an outsider male. She has occupied that same territory with the new male ever since and was nesting at the time of writing. Open wooded areas with plenty of roosting sites and ground cover that contains insects and small mammals, as is the case in this female's territory, have proved to provide for all its needs.

Detailed studies of owlet-nightjar (*Aegotheles cristatus*) habitat found that these birds had home ranges and territories. Home ranges are generally understood as areas that can be visited by various stakeholders of the same species, while territories are exclusively occupied by the territorial owners. Territories may also overlap and lead to tolerated co-foraging in those specific areas.[27] The same may well apply to tawny frogmouths.

Territorial defence

From the few examples actually observed, it seems that territorial competition is strong and appears to be fought out between males only. Tawny frogmouths may seem docile, especially during the day, but once another male enters the territory the resident male acts very swiftly and ferociously. Two examples may suffice here to illustrate. Upon the release of a hand-raised male bird at early dusk (in an area that I thought was free from tawny frogmouth occupancy, incorrectly so as it turned out), a very large male tawny frogmouth appeared, seemingly from nowhere, and raced with tremendous speed towards the released bird, knocking it from behind out of the sky with one massive blow to the back of its head. As the release bird tumbled to the ground, the territorial male followed it very swiftly to the ground, grabbed it by the beak and started shaking it violently, all in a matter of seconds while I was running as fast as I could to the site to save the release bird from a fatal defeat. Even when I finally reached the birds, the wild tawny frogmouth stayed its ground. The objective was obviously to break the neck of the intruder and it would have succeeded had I not picked up the terrified release bird first, separating the two birds by physically handling the beak of the territorial tawny frogmouth. Once the release bird was in my arm, it no longer seemed to constitute a threat and the wild tawny frogmouth flew off.

Beak interlocking between males has also been observed in the Papuan frogmouth in the Lockhart River region on Cape York Peninsula, Queensland.[28]

I have witnessed the beak interlocking fighting technique myself. The resident male and an intruder were locked in battle in the middle of the night, emitting loud screams like cats on the prowl. There was a fierce snapping of beaks, sounding like the slamming of doors, in this struggle of seemingly equally matched adults. Both had their feathers fluffed. When I approached, the intruder flew off. The next day, the damage was more than apparent. The resident male was lying on the ground with several severe slashes to the face and beak. The thin frame of skin on the lower mandible, as described before, was entirely ripped, the tongue hung in tethers and blood was everywhere. The bird was treated and it recovered. I had wondered why tawny frogmouths have such an unusual beak, in that the lower part of the beak is not solid but, as described before, just a layer of stretched skin over the bony frame. It seems now that this is a rather clever construct in that damage to a solid beak might have been far more severe and could have had long-term consequences. The torn section in the lower mandible healed quickly: the skin grew back without any problems and it mended itself amazingly rapidly. The paper-thin tongue also repaired itself. There were some permanent scars but the bird was able to resume normal feeding within three days post-fight and, just as importantly, it maintained its territory.

Defence against birds other than conspecifics, especially diurnal ones, was conducted in an entirely different manner. Either these harassments or attempts to dislodge the tawny frogmouth were ignored, or they were followed by extremely minimalist actions, to surprisingly effective outcomes in most cases. In some instances, young pied butcherbirds were observed to harass a hand-raised tawny frogmouth, perching on a tree for the first time. The frogmouth did not respond to the croaked complaints and constant close fluttering by the butcherbird for a while, then slowly turned its head, opened its eyes fully and raised its hackles. The butcherbird disappeared instantly. A similar response was made in a brief contact with noisy miners (Fig. 5.2). A group of them approached and tried to execute their

Fig. 5.2. A tawny frogmouth nestling and noisy miner juvenile. Curiosity and uncertainty kept the noisy miner looking but not venturing closer.

accomplished mobbing behaviour. Apart from the stoic lack of responding, tawny frogmouths are quite capable of ending the imposition. Just the raising of the head, expanding of feathers and the opening of the beak surprisingly sufficed to send the noisy miners on a swift retreat. Noisy miners are usually not easily intimidated and will mob goshawks and falcons, even eagles, without giving up ground.

Tawny frogmouths can also raise the feathers over their entire body (see Chapter 8, Fig. 8.2). The splashes of dark grey across their plumage now come to the fore, forming a spiky looking crown and making the bird appear much larger and more dangerous than it is. This posture is also seen when approaching a nesting female at night (when fully awake and alert). When I was within a metre from the nest she started swaying and rotating her head while fixating me with her large eyes (see Fig. 6.5 in the next chapter).

Defensive plumage changes have limitations. They may well only work against other native birds and potential avian predators. Cats and rats, as introduced species, do not appear to show similar respect for the plumage displays shown by tawny frogmouths and they may not be an effective weapon against monitor lizards and pythons. For the latter, other strategies are employed.

Tawny frogmouths may very well have the most memorably pungent faeces of any bird and most mammals. Their faeces are not only extraordinarily and overwhelmingly pungent but are also persistent and retain an odour plume until well after they have been expelled. Such a powerful and highly unpleasant odour, one suspects, ought to have a function.

There are two ways in which tawny frogmouths defecate. One is a small stream elimination very much the same as in most other birds. The second method, however, is by delivering a powerful fine and wide-ranging spray. This method seems to be employed as a defensive weapon particularly against snakes and lace monitors, both predatory species that are of great risk to eggs and frogmouth nestlings. Attacks and threats tend to be ineffective against reptiles. However, both reptilian species rely largely on olfaction to pinpoint and follow their prey. The foul spray, delivered by taking aim at their bodies, masks any other smells, overwhelming the reptile and forcing its retreat. It will end their pursuit of tawny frogmouth eggs and nestlings, and hence can be a very effective deterrent. The odour of this substance is also very difficult to reduce, let alone eliminate. It clings to skin and fur, and is even difficult to expunge with strong soap. Tawny frogmouths can indeed be described as 'the skunks of the air'.[29] Since snakes and monitors are very dependent on scent for identifying food, this spraying of faeces is likely to be able to confuse the prowling predator and take away the 'radar' (in this case olfaction) for the successful location of potential prey. It could be surmised that the spraying technique is a strategy that has evolved largely as a nest defence against snakes and lace monitors. It is also possible that possums and other nocturnal native mammals that might try to get too close to a nest site may be warned off by being sprayed with faeces.

In other words, tawny frogmouths are not entirely defenceless and obviously have succeeded in surviving. Among their greatest threats are human activities (ranging from using pesticides to cars at night). Keith Alfred Hindwood noted with approval in 1935 in Sydney that tawny frogmouths fed on centipedes, beetles and spiders. Twelve years earlier, Thomson had published the following observation in *Emu* that is worth quoting and remembering:

> *One cannot but notice the wonderful provision of nature in the adaptation of this bird in a country where insect life is extraordinarily abundant in species and numbers. Probably there is no other bird so entirely useful, from the point of view of man, as the Podargus, nor one which does so much to preserve the 'balance of nature' of which we hear so often. Not only is the Podargus thus one of the economically valuable of all our birds, but it is perfectly adapted for its 'work' – the capture of its insect prey – as well as for its own survival.*[30]

Indeed, it is good to be reminded that many birds (and indeed all native species) are very useful and the tawny frogmouth is a champion. We tend to forget that, without birds, there would probably be many more insect plagues and other problems of greater proportion. Unfortunately, ever larger tracts of land have been clear-felled for farming and other purposes, eliminating the tawny frogmouth from much of its terrain. Retaining or revegetating some wooded areas could serve landholders well. For example, providing tawny frogmouths and owls with a few roosting and nesting trees would result in fewer problems with mice and rats, and reduce or eliminate the need for insecticides and baits, as some experimental cane growers' farms in Queensland have shown.[31]

6

Bonding and breeding

Modern techniques of DNA testing have made all ornithologists very careful not to romanticise the strong bonds of bird pairs and not to be coaxed into believing that the observation of pair bonds necessarily implies faithful relationships at the level of successful matings. We know that, in avian reproduction, even strongly bonded social pairs may raise offspring of which one partner is not the biological parent.[1] We do not know whether this might apply to tawny frogmouths. It seems somewhat doubtful because of their strong territoriality. Evidence suggests that tawny frogmouths form partnerships for life and that such partnerships appear to be rarely contested. Males, as argued earlier, are fiercely territorial and if a strange male should venture into the territory of a tawny frogmouth pair, the intruder is quickly driven off by the local male. One cannot tell, of course, whether the female might leave the territory for brief periods and join neighbouring males or whether some bachelor males may occasionally even succeed in a sneak mating within another's territory. We have to wait until DNA typing of parents and offspring are conducted.

Tawny frogmouths are not cooperative breeders. They form solitary pairs and do not share their territory with friend or neighbour, and they usually evict their offspring in mid-summer, surprisingly early post-fledging (more of this in the next chapter). Pairs tend to stay close together, sometimes perching like boobook owls leaning against each other (Fig. 6.1). We do not know how they form pairs although new research on home range versus territory suggests that the separate home ranges of male and female overlap. It is highly likely that such overlap can lead to pair bonding.[2] There is also evidence that vocalisations may play a role in

Fig. 6.1. A pair of tawny frogmouths on a day roost before the start of the breeding season.

finding a mate and forming partnerships, as well as in maintaining them (see also Chapter 8). We have no reported evidence of a courtship ritual or any hint of male/male competition for a female. However, I have witnessed a male approaching a female during daylight hours and staring at her directly with very dilated pupils (see Fig. 8.4). Pupil dilation as a sexual symbol has been noted in humans and great apes, so it is conceivable that the direct stare with dilated pupils acts as a sexual invitation.[3] Tawny frogmouths express affection by grooming and roosting very close to each other.

Partnerships

No doubt, partnerships in tawny frogmouths are extremely strong. One of the pairs I observed over a six-year period stayed together for the entire time, and this bond ended only when the male was hit by a car and died from his injuries. The female stayed on her own for the next two years even though several males showed up in the territory. Surprisingly, she did not lose the territory but stayed singly in her domain. She ignored all suitors and they eventually left again, until one day she was found building a nest with a particularly handsome and seemingly much younger male, producing two young with the new partner each year for another three seasons when the observation period ended. In another instance, a tawny frogmouth male stayed with his dead partner for four days and was found dead on the fifth. As far as could be ascertained, the male had not moved once in those four days. These are the rare occasions when a question of their ability to grieve certainly comes to mind.

The closeness of the bond may vary throughout the breeding season, and many observers conclude that tawny frogmouths are relatively solitary outside this period. My observations do not support this. Admittedly, it has often taken some time to find the partner once one tawny frogmouth has been located but I have usually found the second one within a 10 to 15 m radius. Clearly, such distances are not classifiable as separations or solitary hunting. While it was often extremely difficult to locate the second tawny frogmouth in the dark, there was evidence that one frogmouth watched the other. Indeed, these attentive preoccupied moments of one of them staring into the darkness often led me to find the second bird.

During the breeding season, their physical distance diminishes considerably. The birds tend to roost together side by side on the same branch, with bodies often touching. It seems to be an integral part of lifelong bonds that physical contact is maintained.

They also display grooming behaviour, a habit that we commonly associate with parrots or primates and a few other long-bonding avian species. Typically, the male will carry out the grooming. It is usually confined to the head region and performed by little snaps of the large beak, which tawny frogmouths use with surprising tenderness. This can be observed in sessions that may last for 10 minutes or more.[4] I have so far only seen males perform active preening and preening of another bird (called allo-preening). The supposedly clumsy beak is employed to gently stroke through the plumage while the female sits completely still, almost as if pretending not to notice.

Tawny frogmouths show a gaping response towards each other like that observed in albatrosses. It is possible that such gaping, together with the grooming and the close roosting, is a strategy to confirm and strengthen the bond. During incubation, one partner roosts on a nearby branch and will also provide food for the brooding partner. Once the young have left the nest the family tends to roost closely together.

Mating

Tawny frogmouths are sexually mature within one year and usually before the next breeding season begins. The only two pairs I have been able to observe from juvenile to mating were a mere eight or nine months old at the time of mating and nest building. Mating in tawny frogmouths is conducted in a slow and circumspect manner. In a pair that I watched mating, the female manoeuvred herself into a comfortable sitting position on a very large branch. There appeared to be no other signal necessary to get the male to approach. No vocalisations were heard, nor did she move her tail, as do currawong and magpie females. Once she was settled, the male flew the short distance to her branch and then simply walked on her back. He gaped and, several times, put his beak around her neck but did not appear to touch it. After nearly three minutes (mating in some birds may take only seconds), he slowly slid off and then settled on the branch next to her. The female later produced two fertile eggs. I have seen matings in the middle of the day and at dusk, so mating does not appear to be confined to a particular time of day or night.

Nest building and the problem of tawny frogmouth nests

Usually, territorial birds need to spend some time, often years, to find a territory before they can reproduce. The same may also happen in the case of tawny frogmouths, of course, but we have no records of this. Given the opportunity, they will mate within the first year of life.

Tawny frogmouths spend considerable time building a nest when they have to start from scratch. Males and females share in the building. The adult building the nest usually brings just a twig or a mouthful of leaves, arrives at the nest's edge and then drops the material from its beak into position. Not much more happens in terms of construction, although both males and females continue to fuss over the nest even after incubation has begun. The circumference varies between nests, depending on the size of the tree fork, but it is at most 30 cm measured diagonally from outer rim to outer rim. The few twigs and sticks that they assemble make a rudimentary nest shape and they do not appear to be arranged in any meaningful way. The loose sticks, usually of 5–6 mm thickness, are piled across each other and the centre, which can be slightly depressed, is softened using leaf litter and grass stems. At best, nests of tawny frogmouths are a precarious affair and they can disintegrate readily. Sometimes, they reuse nests of other species and, where possible, they will choose a large fork in a tree with some near-horizontal branches. If those horizontal branches have any indentations or are surrounded by forks with several upward branches, the actual 'building' of the nest may shrink to a few leaves and sticks with few, if any, tie-downs.[5]

Despite the seeming flaws and even flimsiness, building such a nest may take a surprising amount of time. I have seen a few nests built that took just one night (presumably reused nests) but new ones can take anywhere from one to four weeks before the female can lay her eggs,[6] despite the simple nest structure. Indeed, with the exception perhaps of pigeons, tawny frogmouths may well be regarded as the

Fig. 6.2. This is a perfect nesting tree because of its cup shape and protection from almost all sides. This nest was found in a private garden.

least accomplished among nest-building birds. The failing lies in not tying down the twigs. Any degree of shape and permanence is achieved simply by walking across it, which they seem to do regularly. There are some observations that have indicated that occasionally cobwebs are used as part of the nest and, of course, this is a material that can be used to tie twigs together. How often this might happen during nest building is not known.

One can think of ground-nesting birds that do very little in terms of 'building' a nest. However, the ground itself safeguards the eggs and perfect nests can be made just by scraping the ground or assembling a little leaf litter. Some writers have suggested that tawny frogmouths also used to nest on open ground[7] as many nightjars regularly do, but this is not generally supported in the literature.[8]

Nest building in a tree should withstand rain and storms and accommodate the expected brood up to the time of fledging. Safety is often not well achieved by tawny frogmouths. In the best cases, tawny frogmouths may find trees of a splendid size consisting of several branches rising from a kind of extended platform.[9] Large horizontal forks with branches rising on each side form a natural cup shape and therefore give some protection and support to the nest structure and are probably the ideal nest sites (Fig. 6.2). At the extreme end of unsuitable sites are broken off parts of the trunk of a tree on which eggs are placed.

Perhaps, as long as very old trees with huge branches were readily available, tawny frogmouths never needed to develop nest-building skills to a high degree. It may be a problem to them that very few of these still exist today. The trees available now are generally much younger and smaller, and the branches are thinner. Many indigenous trees have also gone and have been replaced by exotic species that are not suitable nest or foraging sites, if for their relatively steep vertical branches alone. According to one writer, tawny frogmouths prefer, and always choose, native trees[9] and among those prefer very specific kinds. Indeed, a detailed study of 253 tawny nest sites found that most were in rough- or flaky-barked eucalypts; in grassy woodlands preferred species were apple box (33%), Blakely's red gum (31%) and yellow box (26%). In dry sclerophyll forest, they were red stringybark (62%) or a box species (31%), even if dominant trees were smooth-barked eucalypts such as scribbly gums.[10] But tawny frogmouths are not always spoilt for choice. On the entire east coast, particularly, substantial native tree loss may very much limit the offering of suitable nest sites and roosting options. Cryptic colours and manner of roosting had coevolved with native plants, in terms of colouration, tree bark patterns and even the angle of the branches. Horizontal branches tend to appear when a tree is relatively old, hence age of a tree is also important before qualifying as a habitat tree.

Tawny frogmouths do not always build their own nests. They will occupy deserted nests that other species have built, provided they are not located towards the end of a branch, or they will reuse sites. It is possible to see tawny frogmouth nests in deserted nests that were built by currawongs, magpies and white-winged choughs, and relatively unlikely other species, such as black kites and cormorants.[11] In 1930, a rare photo published in Australia's ornithological journal *Emu*, found tawny frogmouth eggs in the deserted nest of white-winged choughs, reprinted here (Fig. 6.3). In 1979, a tawny frogmouth nest was even found in the Nullarbor, the largely treeless desert dividing South Australia from Western Australia, on top of an old cuckoo-shrike nest.[12] In 1928, in Casino, New South Wales, a tawny frogmouth pair was observed nesting for several years on the roots of a large orchid (*Cymbidium*) which grew in a hollow halfway up a gum tree.[13]

Nesting choices may sometimes involve other species. Some species may either act as protectors or feel the need for protection from other birds while some associations are avoided at all cost. At least, it could be predicted that tawny frogmouths would not breed anywhere near ravens. The best known nesting associations in Australian native birds are perhaps that between the noisy friarbird and the leaden flycatcher and between Australian magpies and striated thornbills.[14] The behaviour has also been reported in several avian species worldwide.[15] In frogmouths, one protective nesting association is thought to exist between noisy friarbirds, Australasian figbirds and Papuan frogmouths. It is the only case so far known of a frogmouth acting as a protective species.[16]

Fig. 6.3. Tawny frogmouth nest on an old white-winged chough nest. Source: Dickison D.J. (1930) Unusual nesting sites. *Emu* 30, 145–147.

One of their intrinsic advantages is that tawny frogmouths are nocturnal and very effective in defending against snakes. In snake-rich environments, such as rainforests in the northern part of Australia, there are many nocturnal snake hunters and, indeed, Papuan frogmouths would be excellent guardians against these silent nightly nest robbers. We have no such reports of protective nesting in tawny frogmouths but there has been one observation of friction between tawny frogmouths and the common myna (*Acridotheres tristis*), an introduced species that has already caused substantial harm to hole-nesting parrots. Mynas are known to displace rosellas and other species out of their nesting holes and are ready and capable of even killing the resident small parrot. In a case of role reversal, it was the myna that was being displaced, surprisingly by a tawny frogmouth.[17] So far this seems to be the only report of its kind, so one cannot be certain whether tawny frogmouths would generally discourage other species from breeding near their own nests.

Fig. 6.4. The nests are placed in positions that are well protected from aerial attacks and the brooding tawny (a) is so well camouflaged that a magpie, just landing, is startled (and vocalising) when spotting the tawny frogmouth. The arrow indicates the seated frogmouth with the back turned. Seconds later, the magpie calmed down and did not attack the frogmouth. The position of the nest makes it obvious, however, that the tawny frogmouth is very exposed and vulnerable. The two images show very well that nest building is not a highly developed skill and practice. The 'nest' consists of a few untidy sticks and some leaf litter placed into and around a depression in the tree branch. The arrow in (b) shows that most of the nesting material has already been lost and it hangs down from the branch. While the parents continually protect the nestlings, the nestlings, by that age, more often than not have to contend with the bare substrate.

To come back to the actual nest building, whatever the effort in tawny frogmouth terms, the actual construction has proved time and again to be a serious problem (Fig. 6.4b). When the brooding bird rises from the nest, the sticks move and it is not rare to see smashed eggs on the ground. When the young are hatched, they have to stay well in the centre of the fork or branch because unstable or unsupported twigs will cause their fall from the nest and the tree. The risk increases as the young grow in size and shuffle in the nest themselves, when any movement to the edge usually leads to the helpless nestling falling to its death or sustaining injury. The parents will not usually feed the nestlings on the ground unless they are very near fledging (Fig. 6.5). The fallen nestling will be someone else's food in a matter of a day.

One study[7] suggested that as many as 80 per cent of chicks fall out of the nest and from the tree and meet their end this way. Sometimes parents also accidentally push a youngster off the tree when rearranging themselves on the nest. The current published figures for attrition rate in general is 70 per cent.[18] I am not convinced that we really know enough about their survival rate post-fledging and post-first year to settle on a figure, although there is no doubt that the figure is high.

A further claim has been repeatedly made that loss of nestlings occurs due to storm damage.[19] This may also be a consequence of nest construction; however, we have no current figures indicating how high the attrition rate may be due to nest construction alone. According to statistics kept by wildlife organisations, tawny frogmouth nestlings make a very sad statistic. They regularly produce the highest

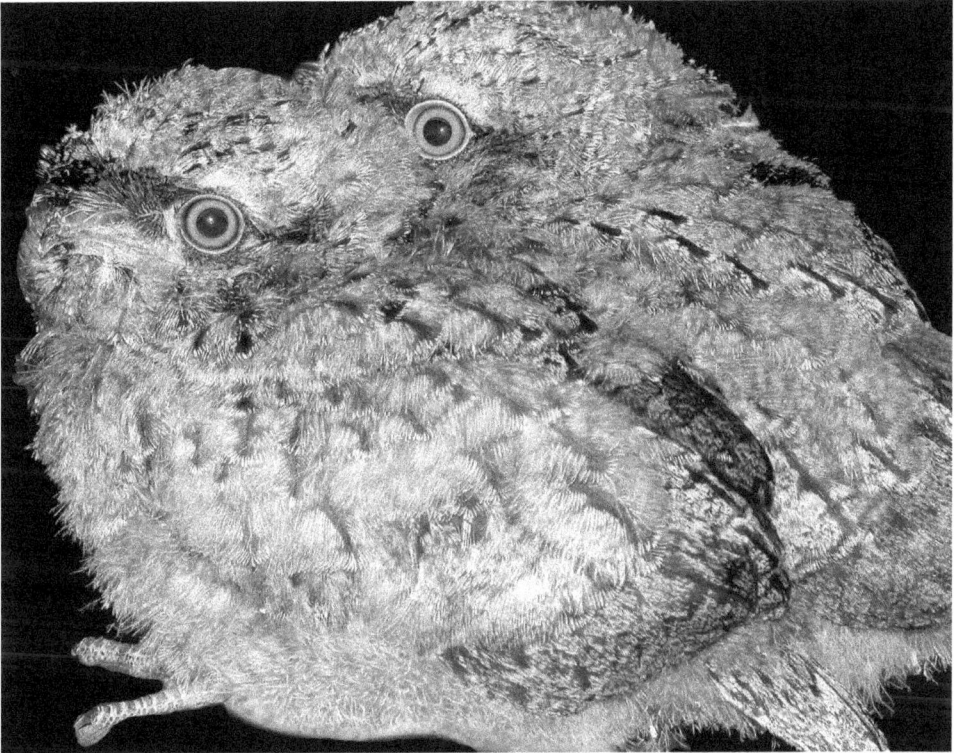

Fig. 6.5. These nestlings on the ground are not near fledging and they would have been doomed if not found. The nest had been destroyed by some mishap so they could not be put back and had to be hand-raised. Surprisingly, even at this young age and with very little wing support, some tawny frogmouth nestlings may fall from great heights (5–10 m) and not suffer any physical (internal or external) injury.

number of orphaned nestlings coming into the care of wildlife organisations in any breeding season. The nestlings that are found, usually on the ground, are the lucky ones. The hand-raising of tawny frogmouths has been a documented practice for over one hundred years and it is part of the reason why these birds have won the hearts of Australians.

However, some more recent studies of tawny frogmouth nest sites[20] have suggested high breeding successes – not all eggs hatched and not all were raised to fledgling stage, but the number of juveniles that were lost was small. Obviously, the success rate cannot be uniformly poor or the species would have serious problems.

Laying the eggs

Once the nest is complete, the female lays her eggs. The eggs are not laid in quick succession, as in some other species. There may be a day, or even two, between the first and the second egg and, again at least a day for the third egg. One study observed that a female took four days to lay a clutch of two eggs.[21] Tawny frogmouth females may lay two or three eggs or possibly even more (although this is unconfirmed) but a large literature on the subject, as well as an extensive nest survey conducted throughout Australia, puts the most common number at two eggs per clutch.[18]

The timing of mating and egg laying has not been studied in tawny frogmouths, although there is plenty of evidence from other bird species that day length (photoperiod) or cues such as rainfall and temperature may trigger breeding behaviour. There has been some evidence in the Sydney area that tawny frogmouths largely breed from early September to October but it may be as late as December.[22]

Parental role in incubating the eggs

In those pairs that I was able to observe at close range, both parents took part in incubation. The male consistently brooded during the daytime and the shift change to the female occurred at late dusk. Several publications have contended that only the female incubates the eggs,[22] but there are as many and authoritative publications that reported that the male sat on the nest during the day.[23] A more recent study using radio-transmitters to examine nesting behaviour showed unequivocally that tawny frogmouths share incubation.[20]

Incubation is a particularly stolid affair in tawny frogmouths. Neither the male nor the female move much at all during their brooding turns on the nest. They just sit. If the weather warms, they may half stand in the nest, presumably to prevent the eggs from getting too warm, but otherwise there is very little of the shuffling and turning of eggs and of brooding adults that one can observe in other species.

In the cases I witnessed, shift change was meticulously observed and quickly executed. When either partner gets ready to leave the nest, the departing adult begins by shaking its plumage and stretching and then will pass a furtive glance at the partner and step off. The waiting bird immediately moves onto the nest. In fact, I have not seen the eggs exposed to the weather for more than a few minutes, even on warm days. Such fast exchanges are necessary in very cold climates but they are somewhat surprising in warmer weather conditions. Bearing in mind that the eggs are large and white, like chicken eggs, they are very conspicuous and one wonders whether the quick exchange is necessary to ensure their safety. Tawny frogmouths, so well known for their camouflage, have eggs without the slightest hint of camouflage. Eggs are turned at shift change at dawn and dusk but not in-between times as has been observed in other species. That tawny frogmouth eggs get turned only once every 12 hours, that is, once per shift of the brooding bird, has also been observed by other writers.[24]

The times for changing shift also appear to be relatively fixed. Males I observed usually did not brood at night and the females never did so during the day.

According to radio-telemetry results of movements on the nest,[20] males and females exchange position several times at night. That too would make sense, allowing both partners to forage. In the case of my own observations, the male departed but then returned to the nest and fed the female at the nest. It may be that feeding practices vary among pairs, or that there may be variations over the incubation period for most pairs. I watched only in the early parts of incubation. On the nights when I stood observing nests throughout the night, the male's absences over the entire night and dawn totalled about two hours, varying between two minutes to 20 minutes in each case, although the latter was rare.

This brooding arrangement may have good adaptive reasons. The first is that the male is slightly stronger and bigger than the female and he may need to stay on the nest during the day for maximum protection of the eggs. Many of the frequently encountered potential diurnal nest raiders, such as ravens and currawongs, are active during the day and this is likely to be the greatest risk period for the eggs. There is little either male or female can actually do against the most common nocturnal predators such as cats or rats except jointly mob the intruder and, in the case of cats, abandon the nest. Second, as the female lays her eggs over several days, the female's preoccupation may make her more vulnerable and she may be better protected by night than during the day in this process. Another possible explanation might be that the male is free to defend the nest at night against any nocturnal intruders without compromising the eggs.

Brooding trance?

Watching brooding in tawny frogmouths reveals surprising behaviour. Males and females appear to react to the world around them in very different ways, suggesting entirely different levels of arousal and alertness. The female tends to respond to the smallest provocation by extending her head feathers out so that it looks as if a large mane adorns the head. The beak is opened a fraction and the feathers of the wing and the back are erected into several layers of fans, almost forming a wheel. Because tawny frogmouths have short lengthwise stripes in their plumage, the entire body appears to be substantially larger and the stripes on the plumage, once the feathers are erect, appear very spiky (see Fig. 6.6b). In other words, the raising of female's plumage takes on the appearance of a threat display, not unlike the display used by frilled-necked lizards. The fanning out of all the head feathers provides the means of appearing larger and more menacing than is the case. The female's level of arousal seems to remain high during the entire time of brooding. Even though the birds pictured in Fig. 6.6 were actually quite tame, the female did not tolerate any close approaches during time of brooding. Her eyes were sharply focused on me and I thought it wise to keep my distance.

The male's behaviour currently defies explanation. The male, pictured in Fig. 6.6a, seems to be in a state of trance, and both males I observed brooding during

Fig. 6.6. An incubating pair. (a) Male incubating during the day; (b) female incubating at night. The female was readily agitated (indicated by the raising of wing and contour feathers and the half-open beak in threat posture) while the male could not be roused to respond at all while incubating. (This pattern changes later in brooding when the youngsters hatch; then there are no apparent behavioural differences between males and females.)

the day showed similar behaviour throughout the entire brooding period (29–31 days). Neither male was reactive to threats or movements and even the familiar and distinct tawny frogmouth camouflage of lengthening the body was barely used. Instead, the male sits completely still on the nest. Most obvious is the strange expression of the eyes: the pupils are constricted and their outline blurred, and they remain so even in dim light (when pupils normally expand). The males did not respond to my approaches and, at times, I almost felt that neither of the males registered my presence. In fact, in order to see whether the eggs had hatched (and in order to take some photos), I gently lifted one of them up and off the nest with both hands and even then, he barely stirred. Similar reports of being able to lift the male off the nest are known.[25]

To my knowledge, there are no examples in the literature that can offer an explanation for this male behaviour. Very clearly, however, male and female adopt very different behaviour during brooding, but post-incubating these differences disappear very quickly.

It would seem quite plausible to hypothesise that brooding males fall into a hormone-induced trance shown in the continuously contracted, ill-defined pupils. Indeed, the same male outside the breeding season, and around the same time of dusk, has typically dilated pupils in the dimming light, the eyes are clear and the edges of the pupil are well defined (Fig. 6.7). It could also be argued that the males did not respond because I had raised them, let alone fly away when approached (as they may do under extreme provocation when a strange human approaches and repeats such approaches several times) but it does not explain the differences in pupil size inside and outside the brooding period.

To my knowledge, this is the first time that these behaviours have been observed and described. The findings raise several new questions. Why would the female show such heightened vigilant behaviour but not the male? What could possibly be the advantage of 'switching off' some attention to the world while being exposed to the elements, sitting motionlessly on a very exposed nest? The description of a chance observation perhaps even deepens the puzzle. This observation made by Steve van Dyck[26] is related here in full. When watching a nest site during the day, he saw the male get up briefly in the nest and that moment was enough to show a waiting crow that there were young chicks in the nest. Van Dyck said that the crow then flew over to the nest and began to pull the tail feathers and the wings of the male vigorously and persistently, eventually unseating the male. The crow then swiftly employed its beak to pull one nestling over the edge, repeating the process until the nest was empty, the brood was lost and the frogmouth pair flew away. Remarkably, the male was absolutely passive and made no single attempt to defend, let alone attack, the invading bird. Clearly, this is

Fig. 6.7. The same male as in Fig. 6.6a photographed outside the brooding period. Note the very different iris and pupil characteristics.

puzzling and open to all sorts of speculations. One wonders whether the stoic 'holding on' tactic might have worked in other contexts. Perhaps the 'trance-like' state merely ensured that he stayed for the duration to help incubate the eggs and brood the young.

One could also argue that greater vigilance would be beneficial for nest protection because the male could fight the nest raider. Tawny frogmouths have been known to attack once the nestlings have hatched, even during the day, but it seems that such attacks may extend only to humans, goannas and carpet snakes – species that have to climb to reach the nest. In one field study, it was reported that ropes, possibly resembling a snake, were attacked when hauled over a branch near the nest.[20] There have been no reported cases of attacks on birds that land near the nest. Despite the bill and the possibility of making impressive threat displays, the tawny frogmouth is at a great disadvantage during the day and can be outmanoeuvred easily by a large and skilful diurnal bird. It would have little chance of succeeding against a crow or raven or, worse, against a bird of prey. Tawny frogmouths do not really have many defensive weapons and if the passive defence of remaining motionlessly on the nest or a bluff of feathers and gaping fails, it appears that there is little that the tawny frogmouth can do.

7

Development

Tawny frogmouths are sometimes thought of as odd birds, and that may not be altogether incorrect, as has already been shown. In their development, tawny frogmouths are also different from other birds and perhaps even unique both in behaviour and appearance.

Indeed, tawny frogmouth nestlings do not behave or look like birds at all. Their juvenile downy feathers look more like fur, their plumage hides the beak, and their plump and 'unbird'-like shape makes them appear like one of the make-believe figures of children's television. Tawny frogmouths undergo several changes during their development from egg to fledging and these stages will now be discussed.

Egg laying timing

Tawny frogmouths belong to the group of birds that lays eggs consecutively, meaning that they are laid days apart. This method of laying occurs in many species but, in some (such as quails), consecutive laying still results in all birds hatching at the same time. However, in tawny frogmouths (as in kookaburras and others) consecutive laying of eggs also results in correspondingly different times of hatching, called 'hatching asynchrony'.[1]

The question is whether tawny frogmouths are altricial or precocial. To pre-empt my findings here: they are an unusual in-between category.

For altricial species, hatching asynchrony is the more characteristic form of producing offspring. It is also more typical for feeding specialists (herbivores,

Fig. 7.1. Hatching asynchrony is well demonstrated here. Two nestlings from the same clutch, having hatched approximately two days apart, show clear developmental and size differences.

insectivores, carnivores, frugivores) than among species that are omnivorous, eating both animals and plants.[2] Tawny frogmouths are specialists, feeding on insects and vertebrates, and thus they display behaviour characteristic of altricial nest-dwelling species.

When hatching has taken place, the tawny frogmouth offspring in the nest may vary considerably in size and stage of development (Fig. 7.1), a fact that only some species take into account when they are feeding the nestlings.[3] In some cases, the variation in hatchling strength and size reduces aggression between siblings and in other cases it increases it. In the latter, as in many birds of prey and sometimes also in kookaburras, siblicide may occur when resources are lean or the younger bird is considerably smaller and weaker. The offspring that hatches first pushes unhatched eggs or the weaker, younger siblings out of the nest or kills them outright.

Fig. 7.2. Four fledged youngsters: oldest far right, 2nd hatched far left, 3rd hatched second from left; the small one, second from right, is the last hatched. Size differences are obvious. The smallest one can barely fly (more like a branchling) but has managed not to fall off the tree. Note not just size but also plumage change differences.

Asynchronous hatching, as is the case in tawny frogmouths, also creates an age hierarchy (Fig. 7.2), so the conditions for one chick to dominate another are present according to the order and timing of egg laying. In kookaburras, some sibling competition occurs early on in development. The smaller ones may be harassed and injured but usually will not die, although deaths have been reported, especially in larger clutches.[4]

If the food supply is limited, competition may become fierce and the weakest may not survive. I have raised nestlings together in the same nest deriving from the same clutch and from different clutches and, in either case, there was no competition among the nestlings. However, all of the hand-raised nestlings were well fed and the degree to which a nestling competes may become more apparent when nestlings are deprived. At least the hand-raising experiences provide some limited evidence that competition between siblings is not strong. On the contrary, in the early parts of development, as nestlings, it appears that tawny frogmouth youngsters support each other by snuggling close together (Figs 7.1 and 7.2).

Egg size

Tawny frogmouth eggs are like small chicken eggs in appearance. They are lustrous white and elliptical (Fig. 7.3). According to *HANZAB* and others,[5] egg sizes vary substantially across Australia: the largest in the south-east are around 51 × 34 mm, in eastern parts of Australia a median value of 43 × 31 mm has been recorded, and the smallest are 37 × 25 mm. There also seem to be variations between subspecies. For instance, the subspecies *Podargus s. rossi* apparently has smaller eggs than *P. strigoides*.[6] However, such variations exist within the same species, let alone subspecies, of most birds.

Egg size (and weight) also depends on the size of the female. Large females lay large eggs and small females lay small eggs and, as a rule across species, for every doubling of adult female weight, egg weight increases by ~70 per cent.[7] Egg size within a species is actually very important: the heavier the egg, the more nutrient reserves are available and the heavier the chick at hatching. Hence, egg weight may ultimately determine the survival of the hatchlings.[8]

As a general rule, the duration of incubation of the eggs is also directly related to egg size. The shortest periods of incubation are found in altricial species that have the smallest eggs. For example, the brown warbler, *Gerygone mouki*, a small bird of 9–11 cm, produces eggs of just 17 × 12 mm for which incubation takes just 12–14 days, but other altricial species may need to incubate for 3–4 weeks and, in exceptional cases as with some parrots, even longer.[9]

By contrast, precocial species, that is, those that are nearly independent when hatching, such as the megapodes (Megapodiidae), typically have long incubation periods. In malleefowl, one of the megapodes, incubation takes ~49 days and the eggs are extremely large, measuring ~92 × 61 mm. In tawny frogmouths, incubation of around 30 days occupies a mid-value between those two extremes, varying between 27 and 31 days,[10] although some writers have put incubation as long as 36 days.[11]

Inside the egg

We know nothing of tawny frogmouth development inside the egg, about its sensory development, the kind and amount of nutrients and other inputs obtained. We know from other avian species that the female may deposit certain amounts of hormones in the egg and this can affect the development of the embryo and the behaviour of the young after hatching. In canaries (*Serinus canaria*), the female deposits increasing amounts of the sex hormone testosterone in the egg, according to order of laying. Canaries hatched from eggs laid later in the clutch grow faster, are more assertive and beg more persistently than those hatched from eggs laid earlier in the clutch;[12] hence, later hatched birds are compensating for their initial disadvantage and body size by growing faster and therefore catching up with siblings hatched earlier. However, this course of events is not the same in all birds. In zebra finches, for instance, the amount of testosterone actually decreases with each egg laid within the clutch.[13] It would be important to know what happens in the case of tawny frogmouths but we know from the nestlings that the size difference remains to fledging. The younger ones usually do not 'catch up' in size to the older ones (illustrated in Fig. 7.2), as is the case in canaries. In some rare instances, synchronous hatching of tawny frogmouths has also been reported[14] but this does not seem to occur often.

As the avian embryo develops, it starts to move inside the egg. At first its movements are uncoordinated and sporadic. Then they become more coordinated and reach a maximum about half way through the incubation period.[15] The wings, legs, head and beak all move and sometimes the whole body is moved and turned. This activity is essential for the developing nerves and muscles.

Each of the sensory systems develops at different times during incubation. From domestic fowl, we know that the sense of touch develops first, at about a quarter of the way through incubation.[16] Next comes the ability to respond to sound, at about half way through incubation, and the embryo moves when it hears sounds of certain pitches. Sensitivity to taste may develop at about the same time as hearing. Just two or three days before hatching, the embryo begins to respond to stimulation by light, meaning that the nerves from the eyes to higher parts of the brain become functional. Finally, the sense of olfaction develops at the stage when the chicken embryo pushes through the egg and starts to breathe air.[17] The ordered sequence of development of the sensory systems from tactile first to olfaction last is very important. It allows the earlier systems to become functional without interference from the later ones. For example, the embryo can hear before it can see and that appears to be essential in determining how the young bird will integrate what it hears and sees after hatching.[18] This detailed knowledge describes the timetable of development in precocial species, such as the chicken. It is possible that some of the developmental stages are delayed or work slightly differently in species that are immature, naked and blind on hatching.

Hatching

Hatchlings generally fall into two groups: either they begin life as helpless nestlings, or they are ready almost immediately post-hatching to walk about and feed themselves. The ones that belong to the helpless category usually stay in the nest and are dependent on care from their parents for all of their needs. They hatch in a very immature state, before they open their eyes and without any feathers (i.e. the altricial pattern of development). All of the passerines, including the small songbirds, fit into this category and so do the owls, eagles and hawks. The young of other avian species hatch at quite an advanced stage of development and can feed themselves. They are referred to as 'precocial'. They hatch with feathers and with their eyes open, and are able to walk very shortly after hatching. Many domestic birds, including chickens, ducks, quail and turkeys, are precocial.

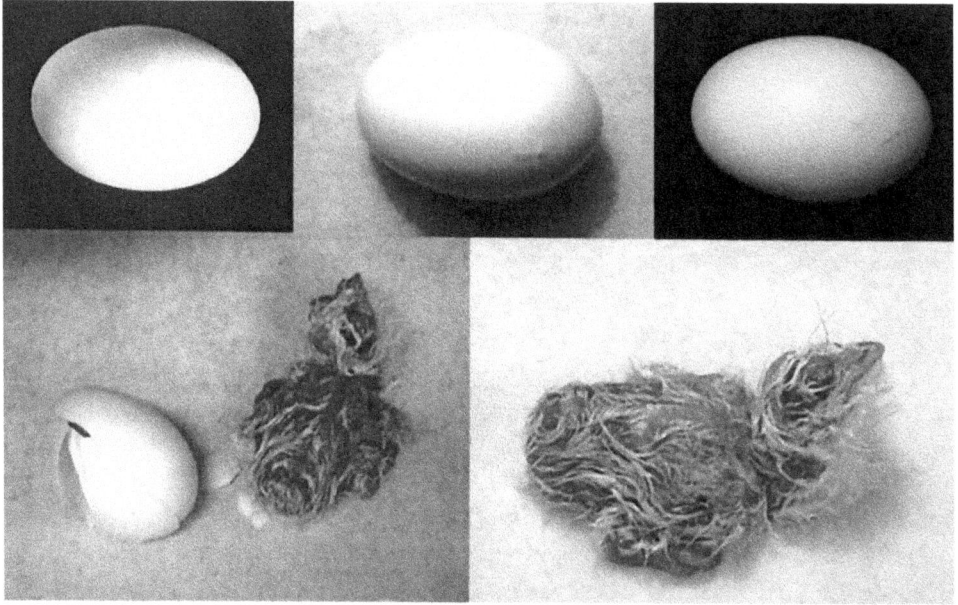

Fig. 7.3. The hatching process. Note the pipped areas in the egg in two different positions along a longitudinal axis. Eventually the egg splits lengthwise, a very unusual way of hatching. The hatchling is already covered in heavy down (unlike other altricial species) and is still wet.

Tawny frogmouths do not quite fall into either category. They are referred to as semi-altricial. That is, they are altricial but with two important differences. While they are helpless at hatching and have their eyes closed, they open their eyes sooner than other altricial species (by about day four[19]) and they have downy feathers from the day of hatching as do precocial species (Fig. 7.3).

Tawny frogmouths, within an hour of hatching, weigh between 17–19 g, based on my own sample ($N = 3$) and that of other studies undertaken on the Northern Tableland of the New England region ($N = 5$).[14]

Hatching is a remarkable event because it requires a large amount of energy and coordinated movements by the embryo. In typical hatching, the embryo proceeds to cut through the blunt, narrow end of the egg and has to manoeuvre itself so that it can first cut the membrane of the air sac and breathe before making its assault on the eggshell itself.

Oddly, and more unusually, tawny frogmouth hatching does not occur from the blunt end of the egg but by splitting the egg in half and lengthwise (Fig. 7.3). It is not clear how the embryo positions itself before hatching and how it obtains air, since the air sac is typically located at the blunt end of the egg. We know of some other species, such as woodcocks *Philohela minor* and sandpipers, family Scolopacidae, in which chicks about to hatch protrude their beaks through the hole pipped in the shell and then also split the shell in a vertical direction (along the longer axis of the egg), instead of chipping around the egg near the air sac.[15]

Tawny frogmouth hatchlings have an egg tooth projecting as a sharp point from the tip of the upper mandible of the beak, as do other species. With the assistance of this egg tooth, the surface of the egg is pipped and the embryo then typically rests before making the final assault on the shell. The hatching embryo enlarges the break in the shell with the beak as it moves its head up and down and in so doing strikes the egg tooth against the shell.[20] When I was observing the hatching process of the tawny frogmouth, it appeared as if there was a thin fault line along the vertical axis of the shell because the areas that were pipped were small but, rather unexpectedly and suddenly, the entire egg fell open into two neat halves and the embryo eventually just shuffled forward and out of the shell.

Nestling stage and camouflage

When the newly hatched bird dries, it is covered in snow-white down, and is only slightly larger than a golf ball (Fig. 7.4). Some of the early stages of development happen in rather quick succession – the white/blind phase is quickly supplanted by eyes opening, upright posture and increasing elements of grey and motley colour interspersing with the white (as Figs 7.4 and 7.5 testify). Within a week to 10 days, as the nestling grows out of 'flower' size, the plumage changes to a mottled off-white or greyish form that blends in well with the nest site in general. Fig. 7.5 shows hatchlings within a day or two of hatching and two 10-day-old nestlings. It is almost impossible to spot the birds. The arrow in the image (B) alerts one to the fact that this is the head of a second nestling that is fully asleep. Hence, it seems that camouflage may be important from the moment of hatching.

One may well wonder why a bird nesting out in the open and subject to predation would produce offspring that are conspicuously snow-white? There may be several, quite different, explanations. White is the simplest feather colour, shared by many raptor offspring, and indicates the absence of pigments. Birds of prey hatchlings are also often snow-white, while most altricial birds grow downy feathers in cryptic colours. Only later may the tail and wing feathers have their species-typical markings and colours.

The white down looks deceivingly like a eucalypt flower and the tawny frogmouth nestlings are about the same size as these flowers (Fig. 7.6). While I thought that this discovery was quite exciting, it could be coincidence rather than adaptive. It is at least worth noting that most eucalypts produce white flowers (the most common of all the flower colours in gum trees) and that there are several *Angophora* and eucalypt species right around Australia that sport large white flowers specifically when tawny frogmouths breed. These have a very similar shape, size and texture to frogmouth nestlings.[21]

And in defence of even considering this thought further, the most obvious tawny frogmouth characteristic is their ability, as juveniles and adults at least, to be

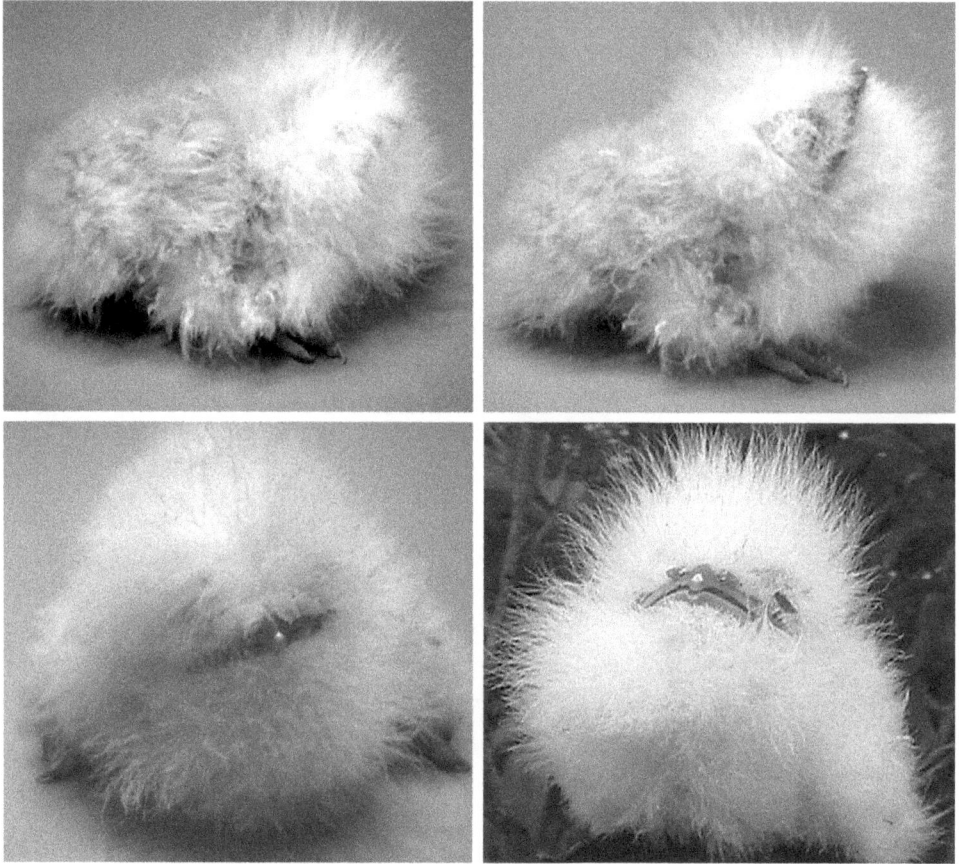

Fig. 7.4. Development in the first four days. The first two images (top) were taken within the first 24 hours, the third (bottom left) on day two and the last image (bottom right) on day four. By then the hatchling shows a very upright posture and the eyes are about to open.

extremely well camouflaged. While hatchlings will be covered by a brooding parent most of the time it is possible to also think of the white down of newly hatched tawny frogmouths as camouflage. The down is not just white but the downy feathers form a tuft over 180° with strands of feathers pointing outwards in a semi-circle, just as gum blossoms have the stamens pointing up and out (see Fig. 7.6). More importantly, any such camouflage makes particular sense only when viewed from the top, be this from the air or from higher branches.

It is worth remembering the debate where tawny frogmouths belong taxonomically. Here, in one very crucial sense at least, tawny frogmouths seem as far removed from owls as can be: owls are predators and at the top of the food-chain. Tawny frogmouths, on the other hand, belong to the large group of birds that are potential prey items, even from owls the size of tawny frogmouths and smaller, and are subject to the same risks as many songbirds and non-songbirds are. Hence, camouflage is certainly a biologically important topic to cover with respect to tawny frogmouths.

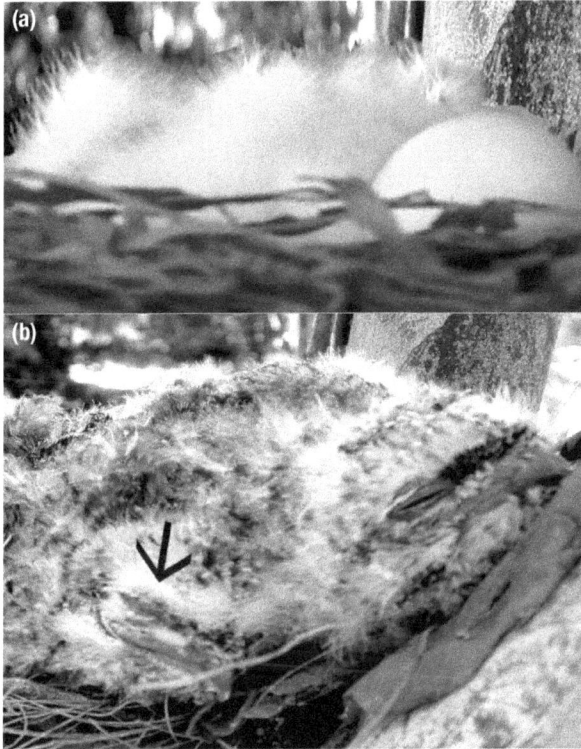

Fig. 7.5. Camouflaged nestlings. (a) Two hatchlings at 24 hours and three days old (the egg did not hatch but, oddly, was not removed by the parents); (b) as the nestlings grow the plumage changes to a mottled off-white and it is difficult to recognise these as birds. The arrow indicates the position of the second nestling.

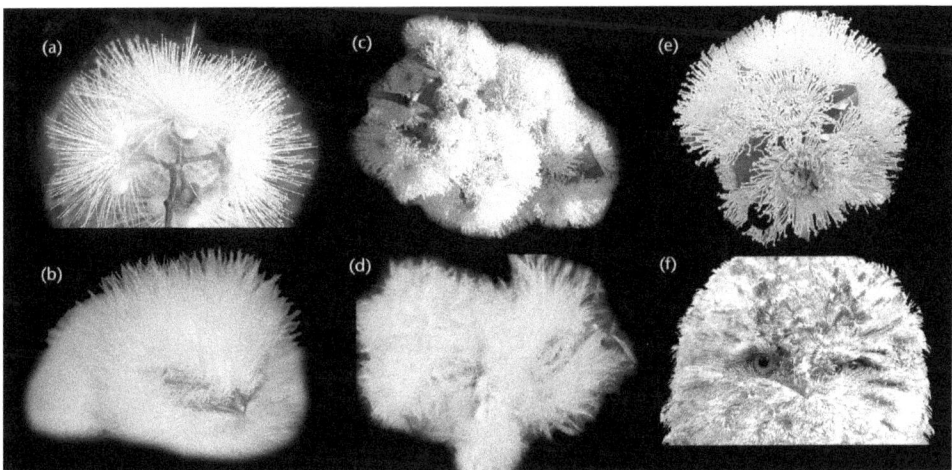

Fig. 7.6. A comparison of gum blossoms (top row) and tawny frogmouth chick plumage (bottom row). Young chicks are a similar size to clusters of gum blossoms and show the same spikes of outward white feathers as the stamens (a,b). When young tawny frogmouths are asleep aged 4–10 days they tend to drop their heads down and support them on the substrate. As a result, all that can be seen from above is a matted jumble of whitish matter resembling tightly clustered gum blossoms (c,d). A frogmouth nestling, with eyes and beak largely hidden in the thick plumage, here resembles a large greyish white cluster of gum blossoms (e,f).

Early development after hatching

Most altricial species will have gained adult weight by the time they fledge, and in some species, such as albatrosses and kookaburras, the fledgling may even be heavier than the adult birds. Tawny frogmouths, by stark contrast, are slow in developing. At the end of the nestling period and at the time of fledging, their weight may just be half of that of adults, having gained weight throughout the nestling period at a rate of 6–10 g per day. No doubt, the exceptionally low metabolic rate of tawny frogmouths in comparison with other birds of similar size may be negatively correlated with growth rate.[14]

In true altricial species, the fast growth of their offspring places enormous demands on the parents, particularly if food is difficult to obtain. Absences from the nest to forage expose the young nestlings to predators and many nestlings are lost in this way. Absences must also be balanced against loss of body heat by the nestlings. Altricial birds generally are unable to maintain their own body temperatures in the early period after hatching; however, it would appear that tawny frogmouth hatchlings are not in the same league as altricial species in general because of their downy cover from day one post-hatching. Thermoregulation is not nearly as much of a problem in tawny frogmouths as in other altricial species. Their thick juvenile plumage protects them extremely well and is in stark contrast to the late and gradual development of plumage in true altricial species.

Downy feathers help the young to stay warm, as does huddling together with siblings and shivering.[22] Keeping the body temperature of the hatchlings within the correct range is a great challenge to most avian parents, particularly those living in either very hot or very cold climates.[23] Dehydration of hatchlings can be a serious problem but, as mentioned before, tawny frogmouth nestlings seem to have partially solved this by minimising water loss through the adaptation of a keratinised hood over the nostrils.

Unlike most altricial species, tawny frogmouth nestlings go through several stages of plumage colouration and feather types: from white to greyish cover, and some adult feather markings on primary flight feathers. Other adult plumage is delayed and develops only about six weeks post-hatching.[24] The length of time nestlings and hatchlings spend sporting their unusual plumage (Fig. 7.7) is probably one of the reasons why the 'cute' look of tawny frogmouths has left such an indelible mark on the Australian psyche. Irby Florence confessed in the ornithological journal *Emu* about a tawny frogmouth nestling:

> *Of all the derelict baby birds we have picked up and cared for until they were fit to rejoin their own folk, this was the most lovable and most gentle. He was more like some soft, confiding kitten or other small mammal than a bird, loving to be nursed and fondled.*[25]

Fig. 7.7. The thick plumage cover of a nestling in a colour still different from the adults (and a good deal thicker) gives the tawny frogmouth nestling its characteristic 'baby' look.

This was written in 1927 and is probably as true a statement of sentiment as in 2017.

Food for the young birds

The first defecation of nestlings occurs after 24 hours[19] and this is the time when the first feeding takes place. The day-old hatchling, in all, is no larger than the beak of a parent bird. Indeed, the offspring would fit comfortably into the beak of either parent. Hence, I was curious to see how an adult could possibly succeed in transferring food to a hatchling. I had especially received some (dead) unfurred mice pups in order to participate in this feeding event and was concerned that even these might have been too large. After depositing the food near the nest to see whether they would use the food and how they would proceed, I waited and watched, a mere half metre away. It took very little persuasion indeed. The male picked up the small mouse, minced it in his beak and then held the end part of the mouse near the nestling, just touching the tip of its tiny, mostly hidden, beak (Fig. 7.8). Bearing in mind that the eyes of the hatchling were still fully closed, that slight

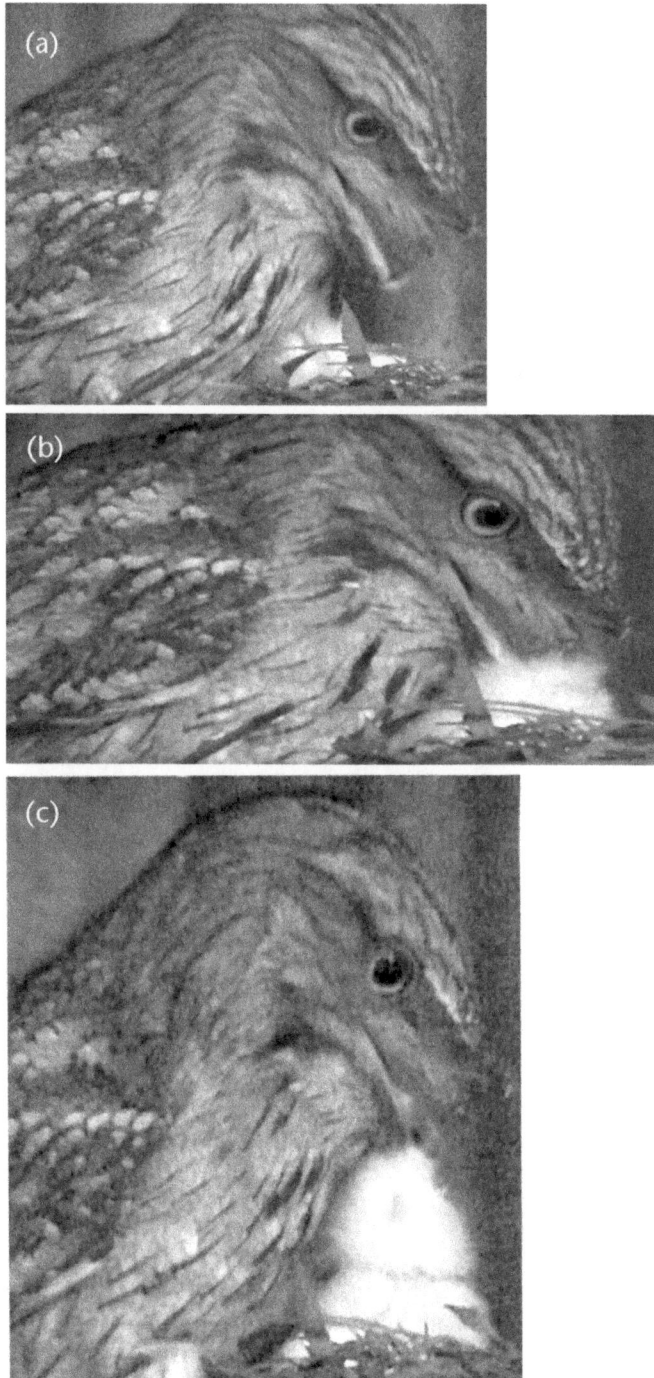

Fig. 7.8. Adult male feeding a one-day-old chick. These are rare photos showing the first feed by the hatchling. The poor quality derives from the fact that it was almost completely dark and no flash was used in order not to disturb the birds. The eyes of the male are heavily converged to focus on the hatchling. The hatchling has its back to the camera.

tap on the beak was all the hatchling received as a signal for food. I now had expected that the bird would gape wide and that the food would simply be dropped into the open beak but this was not the case. Instead, the fragile little bird had to grab hold of the food, and tug on it while the parent bird slowly released the grip on the food, gradually transferring it until the last bits disappeared into the beak of the offspring. Some surprisingly vigorous swallowing followed. The male did not stand in front of but behind the hatchling and his eyes were completely converged to focus on the hatchling in front of him. Every one of his movements was extremely cautious and circumspect and, after the hatchling had fed and slowly lowered its head and collapsed back into an indistinguishable white fluff, the male gently stepped backwards a few centimetres, stood for a few seconds, then turned and flew off the nest while the female stepped back on and covered the nestling with her feathers. All feeds that I observed were conducted in the same manner for the first week post-hatching.

In the weeks following, the parent bird delivering food switched to a position next to the nestling and fed from the side and, by the time they had reached three weeks of age, the nestling faced the adults. Post-fledging, parent birds could barely land on a branch where their offspring were waiting and the food was snatched away almost violently from the beak of the adult. Male and female parents fed equally but, surprisingly, the male was the one to feed the hatchling first. In the weeks post-fledging, I noticed that the female, more than once, had mucus dripping from her beak when she had just fed an offspring. We know that pigeons form what is known as crop milk and regurgitate that to only feed the young. Crop milk is thought to transfer antibodies to the young, assisting in the development of immunity to infection and parasites.[26] Holyoak noted that some nightjars, relatives of frogmouths, feed saliva-enveloped food balls to their chicks.[23] Saliva is sticky and may hold together small insectivorous matter but it may also contain antibodies of importance to the healthy development of nestlings. If this were the case in tawny frogmouths, it would be a very important contribution to the survival of the young. As far as one can tell from the available literature, this remains a rarely reported phenomenon. I managed to take a photo once of the mucus still hanging from the female's beak. It is unfortunately of poor quality, taken at long distance and in poor light, but at least there now is a record of this phenomenon in the tawny frogmouth (Fig. 7.9).

It could also be that the female transfers lubricating fluids to the offspring in the absence of any other source, save for the fluids contained in the food itself. It is perhaps also possible to explain this in another way. As described in Chapter 4, heat-stressed tawny frogmouths engorge the area in the mouth (buccal area) with blood and produce a mucus used for cooling down.[27] Breeding time is stressful for the adults and sex hormone levels rise markedly in this period in females of all avian species. Changing sex hormone levels can alter body temperature but whether this has anything to do with mucus production would be very speculative.

Fig. 7.9. Adult female in the middle and fledgling on the left. The fledgling had just been fed. Light conditions were poor (twilight). The arrows indicate the copious amount of mucus dripping from the beak of the female after feeding her offspring. This phenomenon has never been reported before in this species.

Except for the crop milk in pigeons, it has not been tested as to whether offspring benefit from this mucus and, if so, in what way.

Begging is one of the most noticeable types of behaviour of nestlings of altricial species. They stretch the neck, gape the beak and make begging calls. And here is another surprise: nestling tawny frogmouths barely, if ever, beg. They stretch their necks to collect the food from the parent bird and may utter a few charming and subdued gurgling sounds, but otherwise they seem to be quiet and easily contented. They wait their turn and huddle together regardless of temperature (Fig. 7.2).

However, just before and when they have fledged, the behaviour of young tawny frogmouths seems to go through an utterly obnoxious stage. The charming subdued nestlings turn into demanding monsters and mischievous pranksters. The parent birds are bombarded by loud, and often very prolonged, begging calls that sound more like the screams of a cat in heat than calls made by a bird. The offspring may also pester them, fly towards them, bump them, snap at their head feathers and, additionally, do many things that are risky. For instance, they do not necessarily respect daytime as sleeping time, as said before. The roosting spots to which they fly may be exposed and thus unsuitable for a daytime roost. In one case, a youngster insisted on staying on an iron sheet roof on a hot summer's day. It was sitting out in the open, feet on the tin roof that was heating up. The parents flew back and forth from a good roosting spot to the tin roof, seemingly in an attempt to entice the youngster to a different roosting site. After two hours of trying in broad daylight, the male finally approached the youngster and, walking, shoving and pushing with his chest forced the youngster into a retreat of a slightly bushy area overhanging the roof. There the youngster stayed and the parents had no choice, yet again, but to spend all day roosting in this uncomfortable site in which all the advantages of their natural camouflage were so carelessly flouted.

I watched this particular pair so closely over the entire period of raising their offspring that, in some instances, I began to feel like I was sharing the burden of parenthood. Over the years, the resident tawny frogmouth pair introduced their fledglings to me in every breeding season although they never allowed me to interact directly with them, nor have I ever encouraged this. However, when feeding demands by these fledged and obnoxious youngsters rose to fever pitch, with all the accompanying clamour of screams and angry buttings of the adults, the parent birds often returned to me with a sense of urgency as if demanding that I help with provisioning of their youngsters. In one case, the male even swooped and hit me on the back of the head (quite a thump!) if I was not quick enough to respond and offer some food items. He was pleasant again once his youngsters were fed, and so were the youngsters.

Fledging

To recapitulate here, the breeding cycle in tawny frogmouth takes two months from the start of incubation to fledging: 28–30 days of incubation and a further 27–31 days to fledging.[28] But it is also important to remember that the offspring hatched on different days and that the youngest could be as many as four days younger than the oldest (see Figs 7.1 and 7.2). Despite this developmental difference between youngest and oldest, all nestlings fledge at the same time. Why tawny frogmouth nestlings all fledge together is not clear. This poses a particular problem when parents have successfully reared three or even four offspring. Quite often a clutch produces only one fertile egg, or two at the most, and much of the specific problem of fledging at the same time disappears.

However, if all three or four offspring have been raised successfully, this may have one severe disadvantage, especially for the youngest. The youngest may, in fact, not be quite ready to fly and to leave the nest. It will try but the short flight may end with the youngest fluttering to the ground and not being capable of rising again.[29] Parents will usually continue to feed such a youngster on the ground because, with luck, it may be able to reach some branches above ground within a few hours or days. In those specific cases, parents may sometimes attempt to defend the stranded youngster if a potential predator comes near, something they will not do when a younger nestling, not near fledging, is swept to the ground. On the ground, freshly fledged tawny frogmouths are at great risk and more often than not will not survive.

The short distance flyers and not-so-competent flyers have usually been classified as 'branchlings'[30] before they are proper fledglings. Having watched more than 30 tawny frogmouths fledge, I am convinced that during that first week post-fledging they would probably be more appropriately called branchlings. Tawny frogmouth flights immediately post-fledging can be a funny sight. The flight reminds one of a toddler that just learned to walk – locomotion without a rudder or a stop – and one cannot be quite certain whether there is not an accident waiting to happen. The flight is uncertain and the landings, as in many other newly fledged birds, can be a sorry sight: crash-landing in and through a layer of leaves, over-shooting the branch on which they intended to land or precariously trying to balance when they have managed to aim correctly for a branch.

Although the onset of flying in birds is due to developmental changes that take place as the muscles and nerves mature, perfecting the art of flying requires practice and almost certainly involves learning.

Dispersal of young

In the few cases when it was possible for me to get an accurate timeframe for dispersal, I was surprised how very quickly this happened post-fledging. Three weeks after fledging the parents no longer fed the young and they were not always seen roosting together. There have been reports about post-fledging dependence on parental feeding varying from one to two weeks post-fledging[29] to at least four weeks.[31] The timetable for dispersal does not seem to follow specific patterns for all tawny frogmouths. Some appear to disperse shortly after feeding independence has been achieved while, in some cases, the youngsters appear to be able to stay in the territory close to the next breeding season. Family groups of up to six birds have been seen together in this post-fledging period.[32]

Although I have seen many tawny frogmouths fledge, the actual event of departure is difficult to predict and rare to witness. I have been witness to only two dispersal events (of different clutches and in different years) and in both cases the departure date was no more than four weeks post-fledging. On the first occasion, when the first offspring was about to leave, I heard a soft crying and wailing near the house and went out to investigate. It was midnight. The young tawny frogmouth sat motionlessly on a branch and emitted these sounds continually, changing from intermittent whimpering and loud wailing to sounds that were close to those of a baby crying softly in pain or a cat being hurt. It was really difficult to bear and watch. By 2 a.m. the bird was still doing this and had not moved from its spot. By 5 a.m. the young bird had gone and never returned. On the second occasion, I heard similar whimpering sounds, but the bird departed after only a few hours into the night.

It is not clear whether the parents actively drive their youngsters off their territory or whether the youngsters have an inbuilt clock to know when it is time to leave. It is also not known whether siblings depart together or singly. Regarding the latter, I had the impression that the dispersing siblings do not stay together even for an initial period. It has happened on several occasions that, on release of two birds from the same clutch, one bird was gone while the other still stayed around for a few more days or even weeks. I have not read a single account that has suggested that young tawny frogmouths can stay with their parents and remain in their parent's territory from one breeding season to the next.

Banding records are invaluable but unfortunately only 107 (less than 10%) of a banding project including 1125 birds were ever recovered. Still, they give us important information on dispersal distance from the original banding sites. By far the majority of the recovered tawny frogmouths (87.9%) were found less than 10 km from the original site and only one was found 52 km away from original site. All others had covered distances in between these values.[33] This suggests that tawny frogmouths do not disperse very far. When they disperse, they are facing the greatest dangers and risks yet known to them and it is not clear how many actually survive to the next breeding season. At the time of departure, they are still not even the same weight as their parents.

Another question is how much do tawny frogmouths acquire through learning? From my experience with hand-raising tawny frogmouths, it appears that hunting skills needed to be taught and/or acquired. In each case, be this for insects or larger prey such as frogs, lizards or mice, a parent bird would acquire the morsel and then deliver it to the fledgling while the fledgling watched the process intently. It typically took about two weeks before the fledgling would go down to the ground and try and fetch the item itself but usually only at a time when the parent bird started consuming the prey item itself rather than passing it on to the youngster.

Hawking flights to catch insects at dusk did not seem to require any training by the parents because in the cases I have personally witnessed there was no evidence that either parent had engaged in hawking flights. Yet their youngster performed such hawking flights very expertly within days of fledging – very similar in style to the hawking flights of drongos and fantails. The clumsy little tawny frogmouths that had shuffled along a branch turned out to be extremely agile and competent in the air, elegantly and swiftly moving amongst freshly emerging Christmas beetles, flying around a tree or entering a moth congregation fluttering near a street light.

Tawny frogmouths are thus competent in hunting on the ground, in the canopy and on the wing, although this had been doubted for years. They can find food in relatively good light conditions, dim light or even dark nights and such abilities were displayed not just in a few exceptional cases but even in recently fledged birds, showing a great degree of versatility.

8

Emotions, vocal behaviour and communication

We know very little about the vocal and emotional behaviour of tawny frogmouths. If one were to put all anecdotes and comments together, one might fill just a page or two at best, even when taking into account publications back to 1900 and earlier. I am not certain why this area has been so neglected because tawny frogmouths make many different sounds[1] and they are, in many ways, as expressive emotionally as are parrots. They are not songbirds and do not produce beautiful melodious songs, as do some of Australia's extraordinary songbirds, but they do have an interesting vocal repertoire that obviously meets their needs.

Emotions and autonomic responses

It is not known whether tawny frogmouths have any intentional communication or whether their expressions are part of adaptive affects. Some of the behavioural and physical adjustments that animals must perform to maintain their physiological state can also function as signals. These are known as autonomic responses because they are controlled by the autonomic nervous system. In humans, we know that these responses occur automatically, without conscious control. For example, in cases of extreme fear, the hair on our bodies is raised in preparation for cooling, necessary if we need to flee. Other autonomic responses also occur, but this

Fig. 8.1. Fear response shown by a fledgling. The hackles are raised, indeed, just about all feathers are raised.

particular one is pertinent because raising of the hair is also common in other mammals and raising of the feathers is common in birds. Fluffing of the hair, termed 'piloerection', functions as a signal of fear and, conversely, often acts as a threat display through enlargement of the body size. In young tawny frogmouths, literally all contour feathers of the bird can be raised and this occurs particularly frequently in juveniles. Fig. 8.1 shows that, upon seeing a passing shadow in the sky, this juvenile raised all feathers over the surface of its body, indicating a state of high arousal and possibly even fear. The bird is entirely absorbed in trying to identify what it has seen and its attention is singularly focussed on the flying object. Since the bird was also shaking, a threat display can be ruled out.

Richard Andrew was the first to point out the importance of autonomic responses in displays.[2] Feather raising can be quite difficult to interpret, partly because it also has a role to play in temperature regulation. Slight raising or fluffing of the feathers encloses air around the body and provides insulation against heat loss. Further raising causes ruffling of the feathers (as their tips no longer touch each other) and heat is lost because air is no longer trapped around the body. Ruffling often occurs in aggressive displays and fluffing often occurs when a bird is quiet and submissive. Sleeking of the feathers appears in the aggressive displays of some species. It also appears commonly in states of high arousal and, as such, it has become incorporated into the camouflage posture of tawny frogmouths when a predator or intruder is near (see Chapter 4, Fig. 4.3).

Urination and defecation are other autonomic responses that, not surprisingly, may be used as part of fear or threat displays. As described in detail in Chapter 5, tawny frogmouths often spray their extremely pungent faeces at a predator approaching from below. The autonomic responses have become incorporated into the threat display and a very effective defensive response.

The autonomic nervous system also controls the constriction and dilation of the pupils in the eyes. In Chapter 3, I mentioned briefly that pupil size changes with emotional state in tawny frogmouths. This is assessed by observers even though the ones undergoing the change in pupil size are unlikely to be aware of sending this signal and those receiving it may likewise not realise that they are using this information. This is an aspect of autonomic function that is used involuntarily in communication. The red ring in the eye at the outer edge of the iris in case of rising anger is an extremely parsimonious signal. It involves absolutely no movement by the tawny frogmouth but can be read as a signal not to come any closer.

The size of the pupil varies with emotional state. Humans assess the pupil size of other individuals with whom they are interacting, although they do so quite subconsciously. Pupil size might be an important aspect of communication in animals also. At least we know that it varies with different states of arousal or emotion. Some years ago, it was shown that the pupil size of a parrot constricted whenever the bird produced learned words and also while it was listening to familiar words.[3] There has been no research done to test whether other birds respond to the changes in pupil size but it is possible that they do so.

Emotional states can also be influenced by internal events such as an increase in sex hormone levels, and this can be reflected in signalling. For instance, the amount of hooting in tawny frogmouths may be influenced by the level of sex hormones circulating in the blood. Tawny frogmouths hoot in different patterns during the breeding season. Other hormones (stress hormones) alter internal states and distress calling, which is usually more frequent and louder when birds are more aroused. Hence, several signals may well reflect the emotional state of and physiological processes within the bird and they may or may not be intended for communication with others.[4]

Emotions as visual signals

For a long time, it was considered impossible or unscientific to think of birds expressing emotions beyond autonomic responses, such as the crying of tawny frogmouths after loss of parents or partner as was described here. However, many

Fig. 8.2. A tawny frogmouth raising its wing and back feathers in rising anger. The motion can be accompanied by a low ventricular 'hmm' sound.

emotions have outward expressions and, if they are observed to happen in specific ways in very specific contexts, they become not only observable but also predictable.[5] Whether or not we interpret correctly what we see depends entirely on the context and methodology. If a certain expression keeps reappearing in very specific contexts only or is regularly followed by a specific set of behaviours, one may begin to suspect that there is an internal consistency based on specific internal events which may include thought processes or spontaneous emotions.

Visual communication is used widely among birds, requiring eyesight appropriate for perception of the particular visual signals. Vocal emotional aspects of signalling may also be accompanied by visual information. For example, the position of feathers on the body and head can be altered to give an unambiguous message of displeasure, fear, anxiety, anger or general arousal. There are many different kinds of changes of these feather positions.

Tawny frogmouths and many other avian species have a wide range of body postures available to them for use in visual signalling of a particular message. Head bobbing, arching of the neck, extending the wings outwards and certain sorts of crouching postures may be used in agonistic behaviours.

Signals issued by feather posture alone have rarely been studied systematically, yet they may be quite important within visual range. Many birds fluff their feathers in a certain way when they are ill but they may also raise their feathers as a warning signal.

Tawny frogmouths can raise all their body feathers simultaneously to make them look menacingly larger than they are (Fig. 8.2). This display is not necessarily accompanied by a vocalisation, but it always precedes an attack and it appears to be used predominantly in territorial disputes among conspecifics. I have also seen this particular feather position in a brooding female.

Chapter 6 showed the female brooding on the nest displaying such a warning sign very effectively. All wing feathers were raised, and those on the head and back of the neck formed a semi-circle, and this was even more dramatic as a display because of the black strokes in the plumage. These strokes made the entire outline of the body appear spiky, creating an illusion of a greater arsenal of weaponry than the tawny frogmouth possesses.

The feathers that flank the beak (ear coverts) can be ruffled to express anger and possible attack. Alternatively, the raising of feathers on the back, rather than the wings, can express rising anger and a warning. Usually, as Fig. 8.2 shows, this is accompanied by a slight shift in body posture, head ducking forward as if in a mode of pouncing or readiness for attack. In interspecies interactions, tawny frogmouths tend to 'shrink' their body size by arranging their feathers as tightly as possible to their body and by stretching their necks, as was discussed in detail before (see Fig. 4.4).

Feather extensions or ruffling on the body or head can also express very positive sentiments, and there are many such postures of tenderness that tawny frogmouths

Fig. 8.3. Different 'moods' shown by the same bird. In the left image, it is relaxed but watchful. On the right it shows a typical display of affection. Feathers that are raised in this display are shown by circles marked 1–3. The constricted pupils in both photos merely indicate that both images were obtained during the day (and sunshine).

Fig. 8.4. Sexual attractiveness. A male tawny frogmouth dilates its pupils, distends the feathers at the lower part of the body and the facial feathers as described in Fig. 8.3. These nonvocal gestures, including direct stares, may be considered soliciting signals. Such postures and gestures are followed by mating.

express. I am certain that some will doubt my claims because the tawny frogmouth appears so unimpressed by the world that it can sit through the commotions in a busy veterinarian practice without batting an eyelid, as mentioned in the Preamble. However, same-species interactions fall under different rules.

One of the ways to express affection lies in facial expression. The idea of birds having a 'face' may seem strange because the 'face' has been largely claimed as something belonging especially to non-human primates[5] and to humans. Although it is recognised that many avian species express individuality in their vocalisations, it is also shown in their appearance.

Facial markings are different from one tawny frogmouth to the next and positioning of feathers on the chin or above the beak, on the ear coverts, on top of the head (the crown), and at the nape of the neck independently of the other feathers may provide clues both about its emotional state and possibly about its intentions.

An example is provided in Fig. 8.3. These are two images of the same bird in slightly different light conditions. In the left image, the bird is neutral in expression. No feathers are fluffed and the pupils are contracted in bright sunlight. In the image

Fig. 8.5. Expression of fear, consisting typically of half-open beak, feathers raised on the head and constricted pupils. Unlike the fear response shown in Fig. 8.1 by a fledgling (with all body feathers raised), this expression is more typical of adults. It can change very easily into a threat display.

on the right, the feathers directly under the beak are splayed out and the chin feathers to either side form a kind of outward facing beard. The pupils are much more dilated. How does one know that this is a gentle and positive emotion? This behaviour starts in nestlings towards the end of the third week. It is often accompanied by the nestling moving closer, to touching distance, and often also by vocalisations that only occur in the context of cuddling and such facial feather extensions. These short-distance vocalisations (55–60 dB) are distinct, ring-tone/gurgling sounds. I have found one example in the literature referring to this sound as something sounding like quiet gurgling of nestling curlews.[6] Cuddly and babyish behaviour is often shown by fluffing feathers above and below the beak (see Fig. 4.3) and, as these images demonstrate, they are readily observable in tawny frogmouths.

Adults maintain this call and use it for intraspecific and close communication. Sometimes, expressions of affection are not only expressed by raising the feathers but also in tilting of the head. The feathers of the lower torso may be slightly raised and draped like a dress over the branch on which it is roosting (Fig. 8.4).

Beak signals

Birds have open mouth displays or, rather, open beak displays which, together with other body signals, can be used in fear or threat displays.

Tawny frogmouths use a variety of open beak displays and the inside lining of the large oral cavity is a striking light green colour, effectively displaying the enormous size of the beak and making the bird look more ominous than it actually is.

As a threat, several bird species open their beaks. This may occur without vocalisation but sometimes is associated with hissing or breathing sounds. Like the barn owl, *Tyto alba*, for instance, tawny frogmouths can emit an exhaling sound in warning while the beak is half open, and then sharply clap the beak several times, often without the slightest change in body posture or feather composition. Half-open beaks without any other changes in body posture and feather position can express uncertainty and fear (Fig. 8.5).

For close, conspecific interactions, these facial expressions and beak positions are powerful signals emitted with a minimum of energy expenditure.

Making sounds

The primary sound-producing organ in a bird is the syrinx and the secondary system aiding sound production consists of the larynx, mouth, tongue and laryngeal muscles. Opening and closing of the beak may also affect the song produced.[7]

Birds that have few, if any, syringeal muscles belong to the non-passerines, as do tawny frogmouths. Those that do have at least four pairs of syringeal muscles controlling the syrinx are classified as songbirds.[8] I have autopsied several tawny frogmouths after natural death or trauma and found that they do not appear to have any obvious syringeal muscles (Fig. 8.6) as would be expected. Birds vocalise by expelling air over the elastic membranes of the syrinx housed within the inter-clavicular sac, an air sac in the pleural cavity. In songbirds, the syrinx consists of two parts, one in each bronchus, and each is innervated separately.[9]

Sound is produced by the actions of membranes in the syrinx (see Fig. 8.6). In tawny frogmouths, this syrinx is bronchial and is activated by the actions of exhalation of air but bound and limited by the cartilage surrounding it. The absence of specific syringeal muscles also limits the elasticity and complexity of the membranes and this, in turn, will determine the range of sounds that can be produced and modulated. The air pressure, together with the entire secondary

(a)

Tracheal
Syringeal cartilage

Syrinx

Syringeal muscles

Bronchial
Syringeal cartilage

(b)

Tracheal
Syringeal cartilage

Bronchial
Syringeal cartilage

Primary bronchus

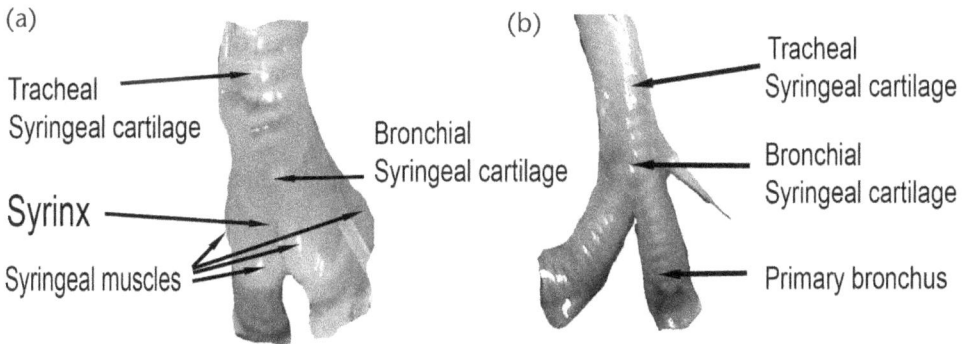

Fig. 8.6.   The primary sound organ of a bird, the syrinx, is located at the point of the bifurcation of the trachea (windpipe) into the two bronchial stems, leading to the lungs. (a) The syrinx of a currawong (*Strepera graculina*) showing strong musculature surrounding the cartilage, identifying this as a songbird. (b) The syrinx of a tawny frogmouth nestling. No syringeal musculature is evident, as is the case in many non-singing birds. Presumably, the trachea will be larger in an adult bird but it is unlikely that any of the musculature would develop later. The tawny frogmouth uses the air sacs for vibration and hence the sounds tend to contain a good deal of noise and appear ventricular.

sound system, still can produce sounds of astonishing variability. The size of the tawny frogmouth, with its long trachea and large body, means that the resonance is generally of low frequency and consists of deep drumming sounds; however, in alarms and threat displays the tawny frogmouth can muster resources of sound of relatively high frequency and very high amplitude that are designed to frighten, even shock, potential predators.

Vocal signals

Vocalisations can signal information about sex, territory or food. They can be used to express anxiety or alarm, rivalry, attention, warnings and similar short instructions. Call types and components have received substantial attention in songbirds over the years, chiefly to establish what their various functions might be.[10] Of course, it is equally important to understand the vocal repertoires of non-songbirds to establish what tawny frogmouth vocalisations are used and in what contexts. Since this work had not been done systematically, I began collecting frogmouth vocalisations some two decades ago and slowly built up a library of calls during different seasons, for young and adult, male and female. In the process, substantial information on the frequency of calls, types of calls, seasonal variations of calls and possible call function has been accumulated and the results of main call types have been summarised in Table 8.1.

The tawny frogmouth uses low amplitude and low frequency sounds to communicate. It has been said that tawny frogmouth vocalisations are barely audible or audible only to 100 m maximum.[11] This takes into account only human

Table 8.1. Vocal repertoire

Whisper communications*	Adult calls	Nestling and juveniles
Purring (like a cat) (nestling–adult)*	Contact oom	Most whisper communications plus:
Crying (soft whimper) (nestling–adult)*	Breeding season oom (staccato)	Begging calls (short and soft) for nestlings, described as quiet gurgling[6]
Hunger (whimper plus accents) (nestling–adult)	Duetting (oom and reply)	Begging calls (strong and long) (juveniles)
Annoyance (sharp little cackles) (juvenile–adult)	Scream (fear, very loud)	Screams (begging) (juveniles, higher freq. than adults)
Courting (bell-like purrs) (adult only)	Scream (threat, loud)	Screams (fear) (juveniles, higher freq. than adults)
Warning 'oom' (long, low hush) (adult only)	Screech (attack/mob – sharp)	
	Alarm (slow cackle)	

*inaudible beyond 1 m

hearing and does not consider what tawny frogmouths themselves might be able to hear. I tested this myself. In the stillness of an Australian bush night, the repetitive hoot of the tawny frogmouth can be heard for at least 500 m. Some of their screams are in the range of 100–125 dB and can be heard for miles. A very simple method to establish whether a transect of bush had a resident tawny frogmouth was to play back hoots of a male. The sounds were played back at 60 dB initially. I then had the idea to reduce the sound amplitude of the hoots and found that wild tawny frogmouths appeared near the sound source as quickly at 35 dB as at 60 dB.

Before and during the breeding season, I have found that males and females often perform duets. Duets can be call sequences that alternate between male and female (that is, only one partner calls at a time) or the calling is simultaneous. Duetting has also been reported in marbled frogmouths.[12] I discovered sexual dimorphism in the dominant call of tawny frogmouths, a hooting call that is generally referred to in the literature as an 'oom',[13] or sometimes as 'drumming'.[14] As it so happens, that is the call they use in duets.

The difference between the male and female calls does not appear great but is quite easy to detect when the birds alternate – the female's hoots are emitted consistently at a higher frequency than the male's. Figure 8.7 shows the deviations of male and female 'ooms' around 1 kHz. For the lowest energy level of the male, around 800 Hz, to the highest of the female, ~1400 Hz, lies a difference of 600 Hz. This is rather substantial and is unlikely to be solely explicable in terms of slight size differences between males and females (body sizes may also overlap). Rather, it seems a specific signal to advertise the sex of caller. When hooting in the dark,

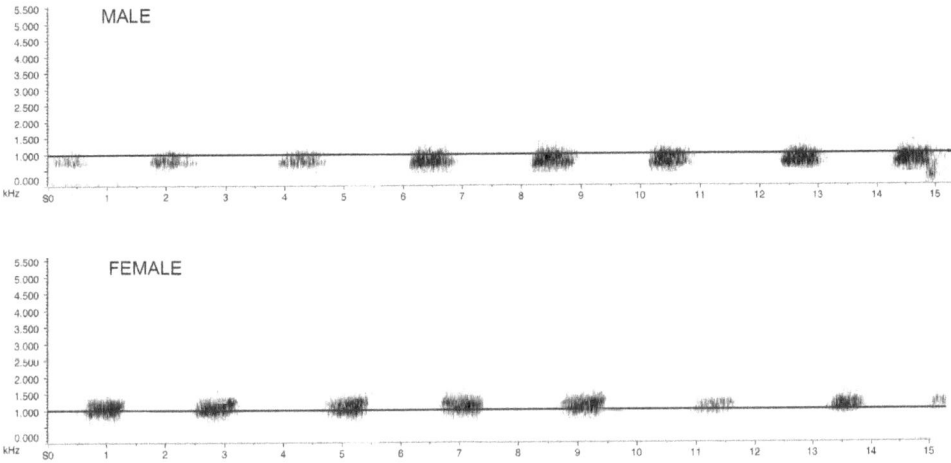

Fig. 8.7. The 'oom' sound by a male and a female. Note that the sound of the male is slightly lower than that of the female, although both calls are around 1 kHz in frequency (*y*-axis). The rate of calling is nearly identical – eight sounds per 16 s (*x*-axis). Darker spots mean greater amplitude (louder call). Both male and female have faint (light grey) and loud (dark grey) 'ooms' that may be produced within the same series of calling.

calls need to be sufficiently differentiated to prevent mistaken identities and potential attacks. The lower the hooting is, the more likely it is a male.

To my knowledge, there is only one account that has described the familiar hoots, or 'oom' sounds, of the tawny frogmouth in more detail and this source is now over 100 years old. In it, Mattingley counted the number of hoots from 14 hoots to 158 without cessation at a rate of six 'ooms' per second.[15] As Fig. 8.7 shows, the hoots I recorded are nowhere near this speed because they are delivered at ~1–2 s intervals per 'oom', delivering about eight 'ooms' per 15 s. These 'ooms' are quite stable, at low amplitude and the rate and speed of interval does not change markedly (with a maximum of 0.2 of a second). Based on many hours of recording samples of these hoots in different locations and over several years, it seems clear that the hoots generally inform about the presence of a tawny frogmouth male in his territory which can be territorial and/or indicating his presence to a female. These soft hoots delivered at low but regular intervals can be heard throughout the year and form the very soundscape of the night.

Hence, the hoots that Arthur Mattingley described in 1910 were actually a different call altogether. However, there is a call that matches Mattingley's description.[15] I have so far heard this only during the breeding season and only produced by the male. It is delivered at the same frequency, sometimes slightly higher (by a maximum of 200 Hz) as the one presented in Fig. 8.7 but it is a good deal faster. This call is indeed delivered at the speed of about six per second. It arises from the steady hoot and suddenly speeds up to a fast, but then steady, staccato (quite similar in properties to the rolling commencement of a kookaburra's 'laughing song') at a rate similar to the human voice being able to say 'ba, ba, ba, ba, ba, ba, ba' at the fastest possible rate.

These fast staccato hoots can be performed with great alacrity. They can be maintained for two to three minutes with only millisecond intervals and can recur several times throughout the night.

There is also a series of soft 'oom' sounds of very low intensity (only ~35–40 dB), although at the same frequency. They are almost hushed sounds, as if we say a low-voiced 'hoohhh' and allow much air to be expelled at the same time. I have heard those sounds when approaching an individual bird. This call appears to be the same call Marshall heard when approaching a tawny frogmouth in the McPherson Ranges, described by the author as, 'a series of soft "cooks" as if in resentment when I approached still closer'.[14] Apart from the sound quality and the low decibel range, this soft 'hooh' call is also well over twice as long as a normal hoot and there are intensity variations even within one call.

Although the three calls (hoots, staccato hoots and low-voiced 'huhh' calls) appear very similar, it is important to realise that most bird species can make very fine discriminatory judgements of auditory signals. For instance, in a detailed study of the acoustic properties of calls in the collared dove (*Streptopelia decaocto*), also not a songbird, variations in rhythm of sounds were found to have important communicative value via the variations of rhythm of territorial cooing.[16] Hoots of owls, as has been found, are indeed important and, among other things, can indicate territorial claims, reproductive success, and even general state of health.[17]

There are several other calls that I have summarised in Table 8.1, including specialised calls by nestlings, begging calls by juveniles and sounds emitted in various stages of hunger. I said before that nestlings do not seem to use begging calls but they do make several calls, possibly expressing distress, annoyance, hunger and fear. Juveniles retain this range of calls for some time, also post-fledging, but add to this the loud scream for begging. These calls are presented in Fig. 8.8.

Affection or affiliative sounds also belong to the repertoire of tawny frogmouths. Such calls made in affection have a distinct set of combinations, as can be seen in Fig. 8.9.

Importantly, tawny frogmouths can also express emotions such as annoyance (Fig. 8.10) and sadness. Annoyance calling is of low amplitude and largely meant for family members. It is used by nestlings, juveniles and adults alike, a cackling that is not so dissimilar to calls emitted by other bird species such as bowerbirds, drongos and many parrot species, also often in similar circumstances, such as squabbles.

Finally, a call (shown in Fig. 8.11) that deserves special comment as it is unusual. I have termed it a 'whimpering' call and it is the closest to crying in an animal that I have yet heard. In fact, there are elements of the call not unlike the low whimper of a newly born human baby in serious pain. It is a gut-wrenching

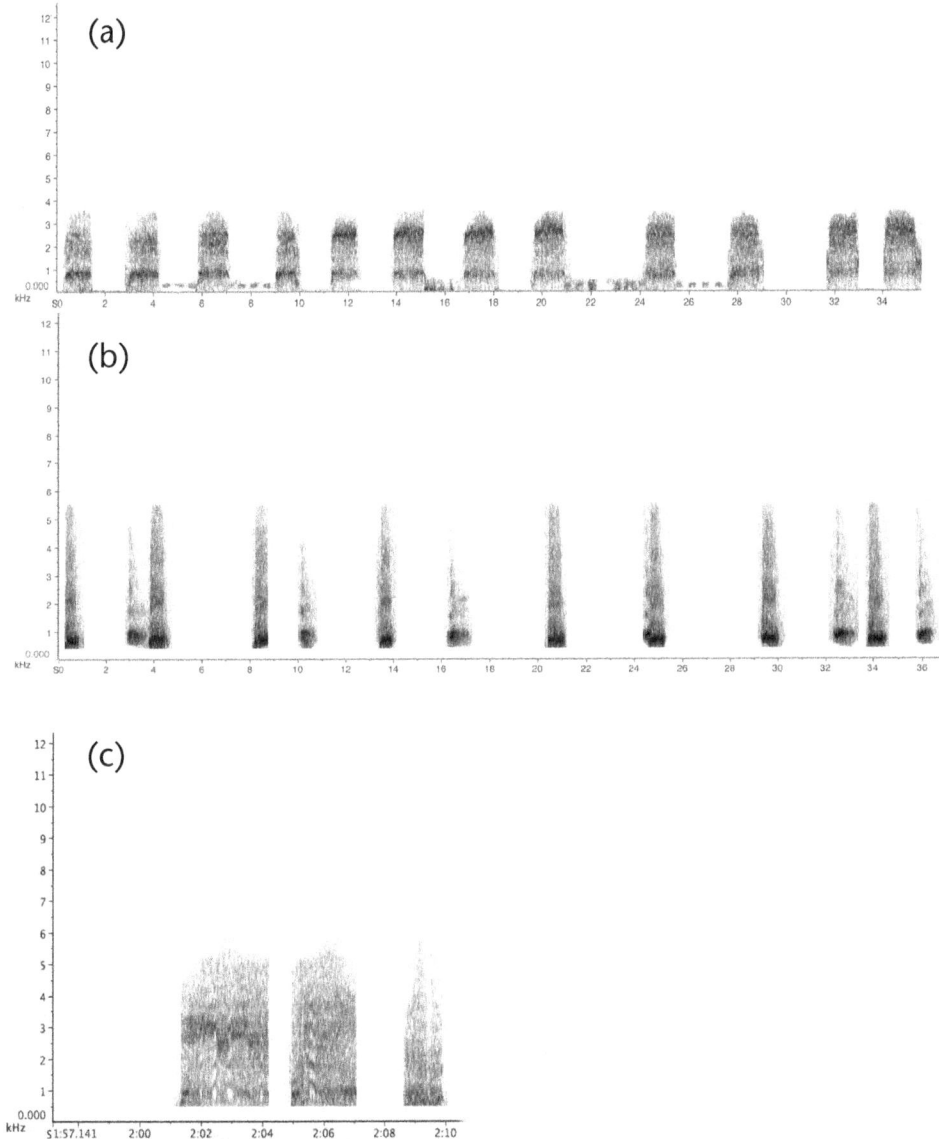

Fig. 8.8. Nestling vocalisations. (a) Sounds made while hungry, but in the absence of adults, that is, they cannot be regarded as begging calls because they are emitted both in the absence of parents and in their presence (before foraging excursions). These calls stop once the chicks are fed. Note the dots along the x-axis – this is the accelerated 'oom' sound by an adult. (b) Intermittent calling performed almost as if coughing. The function is not clear but it could be discomfort. (c) The call of fear is a very loud, noisy call interspersed with some tonal elements. Note 'the ladders' within the call. These are harmonics which make them sound like high-pitched shrieking.

sound and may be emitted at the time when a small nestling tawny frogmouth has lost its parents and, at another time, when the juvenile tawny frogmouth is about to leave the territory. On the night of separation and departure, the dispersing tawny

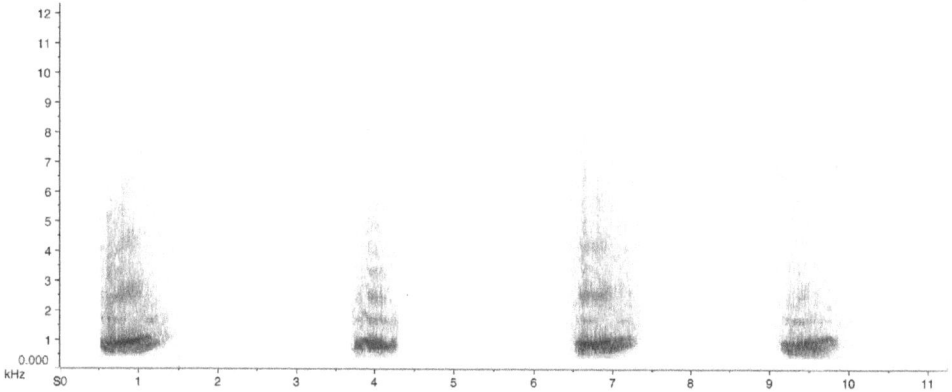

Fig. 8.9. Vocal signal of affection. It is low in intensity but these are surprisingly well-defined calls (note the slight variations) for the tawny frogmouth. The overtones and the energy of the call is around 1 kHz but there is some audible reverberation still at around 2 kHz, so this somewhat faint call (for close interactions only) appears relatively high frequency. It sounds as if quietly gurgling.

frogmouth tends to sit still through the night without feeding and the only sound it makes are these haunting, quiet, whimpering cries. There may be breaks in their expression but, largely, these wailing sounds can continue through most of the night. The largest and most beautiful tawny frogmouth that I had ever raised chose the railing of our veranda as the point of departure and I was able to watch the bird. The calls were incessant and seemed powerfully sad. The bird finally flew away before dawn. The only other context in which I have heard this call is after the resident female had lost her partner in a road accident. She 'cried' for days and I wonder, in this strong pair bond, whether or not her rejection of a gallery of suitors for two years running post-partner death was related to the strength of her bond to her dead partner. Perhaps this indicates an emotion of grieving that was once thought to be unique to humans. The examples have in common that the vocalisations were made in contexts of partner or parent loss. I've heard this in

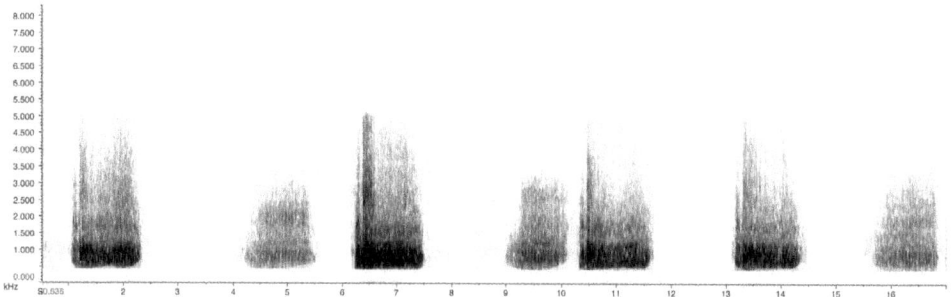

Fig. 8.10. A strong cackling sound of nestlings and also used by juveniles when some annoyance occurs (another sibling pushing, lifting up a nestling). The sound has a characteristic vertical spike at the beginning of each call.

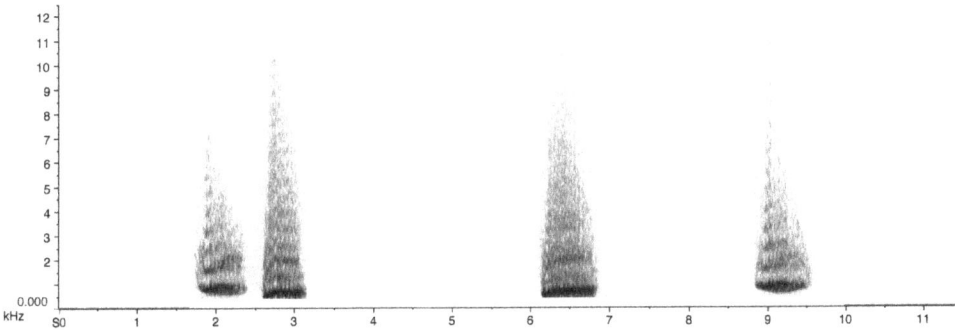

Fig. 8.11. Whimpering/crying by a fledgling at the time of dispersal. A most heart-wrenching sound producing even harmonics and containing bell-like sounds. The calls are very faint (at ~45–50 dB a metre away from its origin) and are the most high-pitched harmonic sounds that I have recorded in tawny frogmouths. It can be performed for many hours (with intervals of a few seconds).

juvenile birds dispersing and in young orphaned tawny frogmouths usually heard repeatedly for several nights post-loss of parents, and in adults after the loss of a life-long partner.

Type of calls

What emerges from the list of calls summarised in Table 8.1 is that tawny frogmouths have quite a number of vocal expressions and, interestingly, a range of loud and faint vocalisations. Faint vocalisations largely seem to be for communication among family members, while loud calls are divided into territorial assertiveness and cross-species communication.

Generalised calls and signals as communication across species

Some of the signals that avian species have evolved appear to function well not only among conspecifics but also across other avian species. They may indeed be so generalised as to be understood by mammals and other vertebrates. Amongst these generalised signals are threat displays, largely visual, and auditory signals such as alarm and distress calls.[18] This may be so partly because of the similar structure of such calls as short, repetitive and loud, and often also because of similarities in frequency of such calls.[19]

The phenomenon of distress and alarm calls has generated a large spate of research and publications. The question is, after all, to what extent a distress call can help or curtail the survival of an individual bird in distress or a bird alarmed at the sight of a predator. Distress calls by tawny frogmouths are not meant so much to communicate with conspecifics, let alone other species, as with the predator itself. Loud screams, even if issued in fear, may be startling enough for the predator to cease the attack or let the bird go, that is, drop it when startled.

Threat displays

Threat displays in tawny frogmouths are extremely elaborate. Signals for the purpose of instilling fear (as a threat) may be issued to conspecifics and against potential predators of entirely different species (reptiles, mammals); hence, these also have to have properties that are not only understood among tawny frogmouths and other birds but also across different classes of animals.

Threat displays aimed at other species, visual and auditory, at times may buy time, that is, they are performed to provide a brief moment during which a potential predator is startled and the bird may have time to get away. Tawny frogmouths use threat displays to bluff their way to safety. Such threat displays, when they cannot be followed by effective ways of attack, may work best when used suddenly and spectacularly, and such displays are certainly spectacular in tawny frogmouths.

In order to achieve this effect, a whole range of simultaneous actions are taken. Half a century ago, Fleay described a case of ruffled head feathers, glaring eyes and extended wings, followed by hollow claps of the beak.[20] Tawny frogmouths may indeed extend all their feathers, including the wings. However, if cornered, they pull in their wings and, in addition, retract their necks fully so that the head rests on their back, almost like the coiling action of a snake. Then they open their beaks very widely so that the entire upper surface of the bird is dominated (and nearly covered) by the enormous gape. Appropriately, this beak action is called gaping. Gaping has been observed in barn owls, owlet nightjars, dollar birds (*Eurystomus orientalis*) and kookaburras (*Dacelo* spp.).[21] The gaping of tawny frogmouths is probably the most dramatic and impressive of all because it fully reveals the enormous size of the inside of the beak. Moreover, the inside of the beak of a tawny frogmouth is of a very unusual colour – a light green, merging at the back of the throat into red. The very small, pale yellow tongue moves, and one's eyes are automatically drawn to the inner red centre of the beak.

If that were not enough, tawny frogmouths emit the most spine-chilling screams, described sometimes as sharp, high intensity calls,[22] that sound like a mixture of the screams of a fighting cat or a pig about to be killed intermingled with some powerful roaring sounds (see Fig. 8.12). Tawny frogmouths will also engage in this display sometimes when they are injured and are picked up by a human. Quite often, though, such actions are not followed by an attack. Presumably, the intruder is meant to take the hint and disappear. If cornered, the recoiled neck can come into action and the head lunges forward with quick, even frenzied, well executed and very powerful snaps of the beak. The edges of the beak are very sharp and the bite-force so substantial that such snaps can inflict injury.

Attack mode is quite different because the bird will move towards an intruder and do so swiftly. This behaviour has not been described in the literature, presumably because it requires long hours of observation and, of course, such

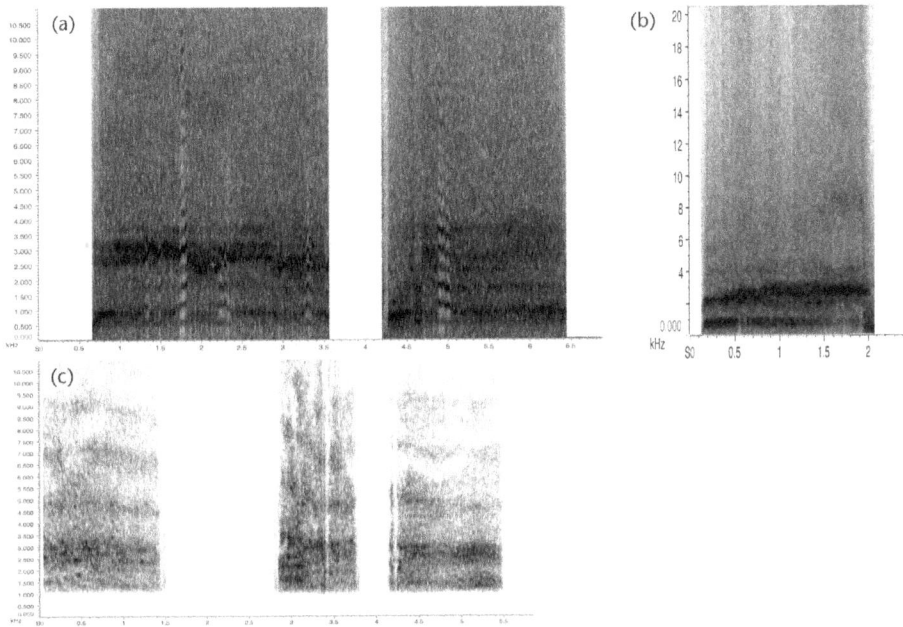

Fig. 8.12. Repertoire of screams. (a) Adult, extreme fear: very noisy and loud call (around 120 dB from 1 m away), main energy around 3 kHz, high-pitched with characteristic feature of embedded 'ladders' (harmonics are very high pitched and sound like the fighting call of a male cat). (b) Scream or screeching sound made at time of displaying mobbing behaviour (note the 'ladders' are missing). (c) Demanding screams by juvenile before and when being fed, may be regarded as begging calls.

events are brief and not predictable. There are several contexts in which I have observed such disputes. In the case of conspecifics (male/male antagonisms), the attack by the resident tawny frogmouth male usually happens silently and from behind, if possible in flight. Having forced the intruder to the ground, a young intruding male may get its neck broken and killed in one swift action. How exactly a tawny frogmouth breaks the neck of another is not quite clear. If in flight, the sheer surprise attack from behind aimed at just below the skull, may be delivered with enough force to achieve this. Alternatively, it is possible that the neck is grasped by the beak and subjected to the same staccato smashing action in which tawny frogmouths smash live prey. That too could break the neck. If the one forced to the ground recovers its wits and faces the opponent, then a battle ensues involving beak locking accompanied by several short screams. I have watched several such interactions. The intruder may end up using the threat display as described above as a last resort and be spared its life and allowed to leave. If the battle is among equally sized and seasoned adult males it can get into a bloody fight with severe injuries sustained to the lower mandible in particular. The skin over the bony frame of the lower mandible is shredded and the tongue may be shredded as well. Usually, those injuries can be overcome and order is restored.

The fights that I have seen have lasted for less than a minute, although beak locking can be maintained for hours.

In case of reptiles approaching a nest site in a tree, the roosting adults may both mob and attack jointly, moving forward and using the beak to hit an approaching snake or goanna at the back of the body, presumably to dislodge the opponent. In these mobbing actions, which they may also employ against human intruders, the parent birds screeched and started swooping, snapping bills.[6] This is a tactic more or less identical to that used by kookaburras with the one exception, as described earlier, that tawny frogmouths can use their excrement to discourage advances by a reptilian predator. The foul excretion confuses any predator relying on olfaction, as snakes and lace monitors do. Moreover, the odour is hard to eliminate, sometimes lasting for days, if the bird manages to deposit some material on the predator's body.

In summary, tawny frogmouths have quite an arsenal of possible signals to express themselves although, of course, that repertoire seems limited in comparison with some non-passerine avian species that live in flocks or extended family groups. Galahs and many members of the Psittacine family have substantially more vocal expressions but not necessarily over the same range. The tawny frogmouth lives in very small groups or on its own. Yet the signals that address risks and dangers are powerful and probably very effective. The signals that deal with emotions are possibly as varied as in parrots and just as endearing and, to the fortunate human who has established a friendship with a tawny frogmouth, irresistibly charming.

Epilogue

In 2005, Karrie Rose commenced her survey on wildlife health with the following words, 'Tawny frogmouths along the east coast of Australia are found moribund, weak, vocalising or convulsing during late winter and early spring of each year'[1] and one begins to appreciate that all is not well. The wild population of tawny frogmouths has to contend not only with its natural enemies and challenges, but also with the ever-increasing number of people in its terrain. And it is not so much the people themselves but what they do, such as clear-felling large tracts of land and using poisons, sprays and general insecticides by the ton that puts tawny frogmouths under threat.

An interesting study of Chicago's birdlife across various regions of the city established that richness of native bird species was inversely related to per capita income – meaning the higher the income bracket, the greater the lack of native birds – while exotic richness (planting non-natives and decorative plants from other continents) was positively related to per capita income. Birdlife was higher in urban sites with undeveloped patches and heterogeneous land cover types, in poorer areas and outer suburbs.[2] These findings could quite easily be translated into Australian capital city conditions. In the last decade or more, there has been a trend for architect designed homes with higher price tags to embrace the 'industrial look' – backyards and front yards filled with tiles, concrete and steel and the few decorative pieces of greenery may be exotic cactus species – as if to ensure that there must be no blade of native grass, soil or even the smallest native shrub growing; a tidy expensive look, a fashion that so unmistakably and aggressively states that any part of nature, let alone any living thing, is banned permanently from these human fortresses. Melbourne had a wave, even a movement, of 'plant natives' in the 1970s and in outer suburbs it allowed the avian wildlife and even smaller terrestrial vertebrates to come back and thrive. Moreover, the city has benefited from cleaner air and a fresh varied look. Green cities are now a goal for some councils but such 'greening' is of little value if the plant species chosen are exotic, as many councils and council-employed gardeners still encourage. Inevitably, such plant cover does not attract the right insects, and does not provide shelter or roosting sites for any native birds. The point is that cities need not be 'foreign cities' by borrowing from other continents – there is more than a small argument to be made that Australian cities would benefit from becoming Australian, with recognisable species of plants that belong to Australia adorning the cityscapes.

It is surprising and pleasing that the tawny frogmouth, because of its adaptability, has fared better than one could have predicted and better than one might have expected given its many vulnerabilities. By comparison, most owls are now scheduled birds, meaning that their numbers are declining, some of them being at higher risk than others. Tawny frogmouths instead, probably also because of their versatility in food consumption, have so far escaped the worst ravages of human-made problems.

Perhaps the tawny frogmouth is a much smarter bird than we thought. It is certainly a tremendously affectionate and curious bird, belying the supposed stoic and disinterested nature often observed. It fooled us with an image of apathy for a long time, but no longer can it hide behind the mask.

This photo was taken when I tip-toed around the tree to get a better angle, trying not to alert the bird to my doings. But then I looked up: the tawny frogmouth followed my every move peering around the trunk of the tree. When I walked around the tree, it peered from the tree in the other direction. They are very curious birds and I suspect do not miss much of what goes on around them.

Endnotes

Chapter 1

1   Van Dyck, S. (2004). Tawdry frogmyths. *Nature Australia* (Winter): 20–21.
2   Ingram, G.J. (1994). Tawny frogmouth. In *Cuckoos, Nightbirds and Kingfishers of Australia*. (Ed. R. Strahan) pp. 88–90. Angus and Robertson, Sydney.
3   Higgins, P.J. (Ed.) (1999). Tawny frogmouths. In *Handbook of Australian, New Zealand and Antarctic Birds. Volume 4: Parrots to Dollarbirds*. pp. 964–983. Oxford University Press, Melbourne.
4   Fleay, D. (1968). *Nightwatchmen of Bush and Plain*. Jacaranda Press, Brisbane.
5   Austin, O.L. (1963). Goatsucker-Caprimulgiformes. In *Birds of the World. A Survey of the 27 orders and 155 families*. pp. 159–163. Golden Press/Hamlyn Publishing House, Felham, Middlesex, England.
6   Barrett, G., Silcocks, A., Barry, S., Cunningham, R. and Poulter, R. (2003). *The New Atlas of Australian Birds*. Birds Australia, Melbourne.
7   Cleere, N., Kratter, A.W., Steadman, D.W., Braun, M.J., Huddleston, C.J., Filardi, C.E. and Dutson, G. (2007). A new genus of frogmouth (Podargidae) from the Solomon Islands – results from a taxonomic review of *Podargus ocellatus inexpectatus* Hartert 1901. *Ibis* 149: 271–286.
8   van Oosterzee, P. (1997). *Where Worlds Collide: The Wallace Line*. Reed Books, Victoria.
9   Ford, J.R. (1986). Avian hybridization and allopatry in the region of the Einasleigh Uplands and Burdekin-Lynd Divide, north-eastern Queensland. *Emu* 86: 87–110.
10 Schodde, R. and Mason, I.J. (1997). Aves (Columbidae to Coraciidae). In *Zoological Catalogue of Australia. Vol. 37.2*. (Eds W.W.K. Houston and A. Wells) CSIRO Publishing, Melbourne.
11 Hollands, D. (1991). *Birds of the Night: Owls, Frogmouths and Nightjars of Australia*. Reed Books, Balgowlah, New South Wales.
12 Kaplan, G. (2015). Australian conditions and their consequences. In *Bird Minds. Cognition and Behaviour in Australian Native Birds*. pp. 1–16. CSIRO Publishing, Melbourne.
13 Hackett, S.J., Kimball, R.T., Reddy, S., Bowie, R.C., Braun, E.L., Braun, M.J., Chojnowski, J.L., Cox, W.A., Han, K.L., Harshman, J. and Huddleston, C.J. (2008). A phylogenomic study of birds reveals their evolutionary history. *Science* 320 (5884): 1763–1768.

14 Gill, F. and Donsker, D. (Eds) (2017). *IOC World Bird List* (v. 7.2). doi :10.14344/IOC.ML.7.2

15 McColl, W.S. (1929). Avifauna of the Hampton Tableland, Hampton Lowlands and Nullabor Plain. *Emu* 29: 91–100; Jones, D. (1986). A survey of the bird community of 'Bullamon Plains' Thallon, Southern Queensland R.A.O.U. Camp-out. *Sunbird* 16 (1): 1–11; Brooker, M.G., Ridpath, M.G., Estbergs, A.J., Bywater, J., Hart, D.S., and Jones, M.S. (1979). Bird observations on the north-western Nullabor Plain and neighbouring regions, 1967–1978. *Emu* 79: 176–190; Cox J.B. and Pedler, L.P. (1977). Birds recorded during three visits to the far north-east of South Australia. *South Australian Ornithologist* 27: 231–250; Hindwood, K.A. and McGill, A.R. (1951). The 'Derra Derra' 1950 Camp-out of the R.A.O.U. *Emu* 50: 217–218; Jones, J. (1951). Victorian Hattah Lakes. *Emu* 52: 225–254; Schodde, R. and Mason, I.J. (1980). *Nocturnal Birds of Australia*. Lansdowne Editions, Melbourne.

16 Woinarski, J.C.Z., Press, A.J. and Russell-Smith, J. (1989). The bird community of a sandstone plateau monsoon forest at Kakadu National Park, Northern Territory. *Emu* 89: 223–231.

17 Hobbs, J.N. (1961). The birds of south-west New South Wales. *Emu* 61: 21–55.

18 Sedgwick, E.H. (1949). Observations on the lower Murchison R.A.O.U. Camp. September 1948. *Emu* 48: 212–242.

19 Hopkins, N. (1948). Birds of Townsville and district. *Emu* 47: 331–347.

20 Haselgrove, P. (1975). Notes on the birds of Groote Eylandt, N.T. *Sunbird* 6: 32–41.

21 Weaving, M.J., White, J.G., Isaac, B. and Cooke, R. (2011). The distribution of three nocturnal bird species across a suburban–forest gradient. *Emu* 111: 52–58.

22 Weaving, M.J., White, J.G., Hower, K., Isaac, B. and Cooke, R. (2014). Sex-biased space-use response to urbanization in an endemic urban adapter. *Landscape and Urban Planning* 130: 73–80.

23 Weaving, M.J., White, J.G., Isaac, B., Rendall, A.R. and Cooke, R. (2016). Adaptation to urban environments promotes high reproductive success in the tawny frogmouth (*Podargus strigoides*), an endemic nocturnal bird species. *Landscape and Urban Planning* 150: 87–95.

24 Cooper, R.M. and McAllan, I.A.W. (1995). *The Birds of Western New South Wales: A Preliminary Atlas*. NSW Bird Atlassers Inc., Albury, NSW.

25 Smith, P. (1984). The forest avifauna near Bega, New South Wales. 1. Differences between forest types. *Emu* 84: 200–210.

26 Gall. B. C. and Longmore, N.W. (1978). Avifauna of the Thredbo Valley, Kosciusko National Park. *Emu* 78: 189–196.

27 Higgins, P.J. (Ed.) (1999). Tawny Frogmouths. In *Handbook of Australian, New Zealand and Antarctic Birds. Volume 4: Parrots to Dollarbirds*. p. 967 [more surveys are listed]. Oxford University Press, Melbourne.

28 Loyn, R.H. (1985). Bird populations in successional forests of Mountain Ash *Eucalyptus regans* in Central Victoria. *Emu* 85 (4): 213–230.

29 Saunders, D.A. and Ingram, J. (1995). *Birds of Southwestern Australia*. Surrey Beatty and Sons, Chipping Norton, Australia.

Chapter 2

1 King, A.S. and McLelland, J. (1984). *Birds: Their Structure and Function*. 2nd edn. Balliere Tindall, Sussex England.

[2]   Higgins, P.J. (Ed.) (1999). Tawny frogmouths. In *Handbook of Australian, New Zealand and Antarctic Birds. Volume 4: Parrots to Dollarbirds*. pp. 964–983. Oxford University Press, Melbourne.

[3]   Rose, K. (2005). Common Diseases of Urban Wildlife. Birds, Part 2. Report by The Australian Registry of Wildlife Health. Co-sponsored by the Western Plains Zoo, Dubbo and Taronga Zoo, Sydney. p. 1.

[4]   Fleay, D. (1968). *Nightwatchmen of Bush and Plain*. Jacaranda Press, Brisbane, Australia.

[5]   For a detailed description on how birds can move their facial feathers including diagrams see Kaplan, G. (2015). *Bird Minds: Cognition and Behaviour in Australian Native Birds*. CSIRO Publishing, Melbourne.

Chapter 3

[1]   Wallman, J. and Pettigrew, J.D. (1985). Conjugate and disjunctive saccades in two avian species with contrasting oculomotor strategies. *Journal of Neuroscience* 5 (6): 1418–1428.

[2]   Iwaniuk, A.N. and Wylie, D.R.W. (2006). The evolution of stereopsis and the Wulst in caprimulgiform birds: a comparative analysis. *Journal of Comparative Physiology A* 192: 1313–1326.

[3]   Wild, J.M., Kubke, M.F. and Peña, J.L. (2008). A pathway for predation in the brain of the barn owl (*Tyto alba*). *Journal of Comparative Neurology* 509 (2): 156–166.

[4]   Schaeffel, F., Howland, H.C. and Farkas, L. (1986). Natural accommodation in the growing chicken. *Vision Research* 26: 1977–1993.

[5]   Martin, G.R. and Katzir, G. (1999). Visual fields in short-toed eagles, *Ciraetus gallicus* (Accipitridae), and the function of binocularity in birds. *Brain, Behavior and Evolution* 53: 55–66.

[6]   Pettigrew, J.D. and Konishi, M. (1976). Neurones selective for orientation and binocular disparity in the visual Wulst of the barn owl (*Tyto alba*). *Science* 193: 675–678.

[7]   McFadden, S.A. (1993). Constructing the three-dimensional image. In *Vision, Brain, and Behavior in Birds*. (Eds H.P. Zeigler and H.-J. Bischof) pp. 47–61. The MIT Press, Cambridge, Massachusetts.

[8]   Vanderwilligen, R.F., Frost, B.J. and Wagner, H. (1998). Stereoscopic depth perception in the owl. *Neuroreport* 9: 1233–1237.

[9]   Miles, F.A. (1998). The neural processing of 3-D visual information: evidence from eye movements. *European Journal of Neuroscience* 10: 811–822.

[10] Menzel, C.R. and Menzel, E.W.J. (1980). Head-cocking and visual exploration in marmosets (*Saguinus fuscicollis*). *Behaviour* 75: 219–233; Rogers, L.J., Stafford, D. and Ward, J.P. (1993). Head cocking in galagos. *Animal Behaviour* 45: 943–952; Kaplan, G. and Rogers, L.J. (2006). Head-cocking as a form of exploration in marmosets and its development. *Journal of Developmental Psychobiology* 48 (2): 551–560.

[11] Payne, R.S. (1971). Acoustic location of prey by barn owls (*Tyto alba*). *Journal of Experimental Biology* 54: 535–573; Martin, G.R. (1990). *Birds by Night*. T. and A.D. Poyser, London.

[12] cf. Kaplan, G. and Rogers, L.J. (2006). Head-cocking as a form of exploration in marmosets and its development. *Journal of Developmental Psychobiology* 48 (2): 551–560.

[13] Martin, G.R. (1990). *Birds by Night*. T and A.D. Poyser, London.

[14] King, A.S. and McLelland, J. (1984). *Birds: Their Structure and Function*. 2nd edn. Balliere Tindall, Sussex, England.

15   Martin, G.R. (1985). Eye. In *Form and Function in Birds. Vol. 3*. (Eds A.S. King and J. McLelland) pp. 311–373. Academic Press, London.

16   Rogers, L.J. and Kaplan, G. (1998). *Not Only Roars and Rituals: Communication in Animals*. Allen and Unwin, Sydney, Australia.

17   Morris, D. (1980). *Manwatching. A Field Guide to Human Behaviour*. Triad Granada, imprint of Jonathan Cape, London.

18   Rojas, L.M., McNeil, R., Cabana, T. and Lachaoelle, P. (1997). Diurnal and nocturnal visual function in two tactile foraging waterbirds: the American White Ibis and the Black Skimmer. *The Condor* 99: 191–200.

19   Kühne, R. and Lewis, B. (1985). External and middle ears. In *Form and Function in Birds* (Eds A.S. King and J. McLelland) pp. 227–272. Academic Press, London.

20   Payne, R.S. (1962). How the barn owl locates prey by hearing. *Living Bird* 1: 151–159; Payne, R.S. (1971). Acoustic location of prey by barn owls (*Tyto alba*). *Journal of Experimental Biology* 54: 535–573.

21   Mason, J. and Clark, L. (1995). Capsaicin detection in trained European starlings: the importance of olfaction and trigeminal chemoreception. *Wilson Bulletin* 107, 165.

22   Bang, B.G. (1971). Functional anatomy of the olfactory system in 23 orders of birds. *Acta Anatomica* 79: 1–76.

23   Grubb, T.C. (1972). Smell and foraging in shearwaters and petrels. *Nature* 237: 404–405.

24   Jones, R.B. and Roper, T.J. (1997). Olfaction in the domestic fowl: a critical review. *Physiology and Behavior* 62: 1009–1018.

25   Nef, S., Allaman, I., Fiumelli, H., De Castro, E. and Nef, P. (1996). Olfaction in birds: differential embryonic expression of nine putative odorant receptor genes in the avian olfactory system. *Mechanisms of Development* 55: 65–77.

26   Clark, L. and Mason, J.R. (1987). Olfactory discrimination of plant volatiles by the European starling. *Animal Behaviour* 35: 227–235.

27   Goujon, E. (1869). *Sur un appareil de corpuscules tactiles situe dans le bec des perroquets*. E. Martinet.

28   Gottschaldt, K.-M. (1985). Structure and function of avian somatosensory receptors. In *Form and Function in Birds, Vol. 3*. (Eds A.S. King and J. McLelland) pp. 375–461. Academic Press, London, New York.

Chapter 4

1   Higgins, P.J. (Ed.) (1999). Tawny frogmouths. In *Handbook of Australian, New Zealand and Antarctic Birds. Volume 4: Parrots to Dollarbirds*. pp. 964–983. Oxford University Press, Melbourne.

2   Kaplan, G. (2004). *Australian Magpie: Biology and Behaviour of an Unusual Songbird*. CSIRO Publishing, Melbourne.

3   Frauca, H. (1973). Tawny frogmouth. *Australian Birdlife* 1 (2): 19–22.

4   Lord, E.A.R. (1956). The birds of the Murphy's Creek District, southern Queensland. *Emu* 56: 100–128; Dowling, B., Seebeck, J.H. and Lowe, K.W. (1994). Technical Report. Arthur Rylah Institute of Environmental Research, Heidelberg, Victoria. p. 34; Dickman, C.R. (2009). House cats as predators in the Australian environment: impacts and management. *Human–Wildlife Conflicts* 3 (1): 41–48.

5   On a 300 km stretch of road between Canberra and Lake Cowal, by far the largest number of birds killed were magpies, followed by galahs, magpie larks and eastern rosellas. Tawny frogmouths were well down that list, accounting for about 2 per cent of all avian road kills collected. Vestjens, W.J.M. (1973). Wildlife mortality on a road in New South Wales. *Emu* 73: 107–112.

6   Rose, A.B. and Eldridge, R.H. (1997). Diet of the tawny frogmouth *Podargus strigoides* in eastern New South Wales. *Australian Bird Watcher* 17: 25–33.

7   Weaving, M.J., White, J.G., Isaac, B. and Cooke, R. (2011). The distribution of three nocturnal bird species across a suburban–forest gradient. *Emu* 111: 52–58.

8   Fleay, D. (1968). *Nightwatchmen of Bush and Plain*. Jacaranda Press, Brisbane, Australia.

9   Körtner, G. and Geiser, F. (1999). Roosting behaviour of the tawny frogmouth (*Podargus strigoides*). *Journal of Zoology, London* 248: 501–507.

10 Schodde, R. and Mason, I.J. (1980). *Nocturnal Birds of Australia*. Lansdowne Editions, Melbourne.

11 Rattenborg, N.C., Lima, S.L. and Amlaner, C.J. (1999). Facultative control of avian unihemispheric sleep under risk of predation. *Behavioural Brain Research* 105: 163–172; Rattenborg, N.C., Amlaner, C.J. and Lima, S.L. (2000). Behavioral, neurophysiological and evolutionary perspectives on unihemispheric sleep. *Neuroscience and Biobehavioral Reviews* 24: 817–842.

12 Rattenborg, N.C., Amlaner, C.J. and Lima, S.L. (2000). Behavioral, neurophysiological and evolutionary perspectives on unihemispheric sleep. *Neuroscience and Biobehavioral Reviews* 24: 817–842.

13 Lasiewski, R.C. and Bartholomew, G.A. (1966). Evaporative cooling in the poorwill and the tawny frogmouth. *Condor* 68: 253–262.

14 See Lasiewski, R.C., Dawson, W. R. and Bartholomew, G.A. (1970). Temperature regulation in the little Papuan frogmouth, *Podargus ocellatus*. *Condor* 72: 332–338; Bennet, P.M. and Harvey, P.H. (1987). Active and resting metabolism in birds: allometry, phylogeny and ecology. *Journal of Zoology, London* 213: 327–363; McNab, B.K. and Bonaccorso, F.J. (1995). The energetics of Australian swifts, frogmouths, and nightjars. *Physiological Zoology* 68: 245–261; Bech, C. and Nicol, S.C. (1999). Thermoregulation and ventilation in the tawny frogmouth, *Podargus strigoides*: a low-metabolic avian species. *Australian Journal of Zoology* 47: 143–153; McKechnie, A.E. and Mzilikazi, N. (2011). Heterothermy in Afrotropical mammals and birds: a review. *Integrative and Comparative Biology* 51 (3): 349–363.

15 Normally 19–20 breaths per minute in thermoneutral environments, at ambient temperature of about 20°C and resting body temperature of 37.6–38.2°. Bech, C. and Nicol, S.C. (1999). Thermoregulation and ventilation in the tawny frogmouth, *Podargus strigoides*: a low-metabolic avian species. *Australian Journal of Zoology* 47: 143–153.

16 Reinertsen, R.E. (1983). Nocturnal hypothermia and its energetic significance for small birds living in the arctic and subarctic regions. A review. *Polar Research* 1: 269–284.

17 Körtner, G., Brigham, R.M. and Geiser, F. (2001). Torpor in free-ranging frogmouths (*Podargus strigoides*). *Physical and Biochemical Zoology* 74 (6): 789–797.

18 Geiser, F. (2012). Hibernation. *Current Biology* 23 (5): 188–193.

19 Doucette, L.I., Brigham, R.M., Pavey, C.R. and Geiser, F. (2012). Prey availability affects daily torpor by free-ranging Australian owlet-nightjars (*Aegotheles cristatus*). *Oecologia* 169: 361–372.

20 Rae, S. and Rae, D. (2013). Orientation of tawny frogmouth (*Podargus strigoides*) nests and their position on branches optimises thermoregulation and cryptic concealment. *Australian Journal of Zoology* 61 (6): 469–474.

Chapter 5

[1] Frauca, H. (1973). Tawny frogmouth. *Australian Birdlife* 1 (2): 19–22.

[2] Hollands, D. (1991). *Birds of the Night: Owls, Frogmouths and Nightjars of Australia.* Reed Books, Balgowlah, New South Wales.

[3] McCulloch, E.M. (1975). Variations in the mass of captive tawny frogmouths. *The Australian Bird Bander* 13 (1): 9–11; Rose, A.B. and Eldridge, R.H. (1997). Diet of the tawny frogmouth *Podargus strigoides* in eastern New South Wales. *Australian Bird Watcher* 17: 25–33.

[4] Jetz, W., Steffen, J. and Linsenmair, K.E. (2003). Effects of light and prey availability on nocturnal, lunar and seasonal activity of tropical nightjars. *Oikos* 103: 627–639.

[5] Davey, C. and Einoder, L. (2001). Predation of house mice by the tawny frogmouth *Podargus strigoides*. *Australian Bird Watcher* 19 (3): 103–104.

[6] Barker, R.D. and Vestjens, W.J.M. (1979). *The Food of Australian Birds 1. Non-Passerines.* CSIRO, Melbourne.

[7] Rose, A.B. and Eldridge, R.H. (1997). Diet of the tawny frogmouth *Podargus strigoides* in eastern New South Wales. *Australian Bird Watcher* 17: 25–33.

[8] Fleay, D. (1968). *Nightwatchmen of Bush and Plain.* Jacaranda Press, Brisbane, Australia.

[9] Ridley, E. (1985). A tale of a tawny frogmouth. *Bird Keeping in Australia* 28: 144–145.

[10] Fleay, D. (1968). *Nightwatchmen of Bush and Plain.* Jacaranda Press, Brisbane, Australia; Ridley, E. (1985). A tale of a tawny frogmouth. *Bird Keeping in Australia* 28: 144–145; Ingram, G.J. (1994). Tawny frogmouth. In *Cuckoos, Nightbirds and Kingfishers of Australia.* (Ed. R. Strahan) pp. 88–90. Angus and Robertson, Sydney.

[11] Schodde, R. and Mason, I.J. (1980). *Nocturnal Birds of Australia.* Lansdowne Editions, Melbourne; Rose, A.B. and Eldridge, R.H. (1997). Diet of the tawny frogmouth *Podargus strigoides* in eastern New South Wales. *Australian Bird Watcher* 17: 25–33; Higgins, P.J. (Ed.) (1999). Tawny frogmouths. In *Handbook of Australian, New Zealand and Antarctic Birds. Volume 4: Parrots to Dollarbirds.* pp. 964–983. Oxford University Press, Melbourne.

[12] Coleman, E. (1946). Foods of the tawny frogmouth. *The Victorian Naturalist* 63: 111–115.

[13] Turner, J.R. (1992). Effect of wildfire on birds at Weddin Mountain, New South Wales. *Corella* 16 (3): 65–74.

[14] Haselgrove, P. (1975). Notes on the birds of Groote Eylandt, N.T. *Sunbird* 6: 32–41.

[15] Ingram, G.J. (1994). Tawny frogmouth. In *Cuckoos, Nightbirds and Kingfishers of Australia.* (Ed. R. Strahan) pp. 88–90. Angus and Robertson, Sydney.

[16] Schodde, R. and Mason, I.J. (1980). *Nocturnal Birds of Australia.* Lansdowne Editions, Melbourne.

[17] Kaplan, G. and Rogers, L.J. (2013). Stability of referential signalling across time and locations: testing alarm calls of Australian magpies (*Gymnorhina tibicen*) in urban and rural Australia and in Fiji. *PeerJ* 1: e112. doi:10.7717/peerj.112

[18] Charles, J.A. (1995). Organochlorine toxicity in tawny frogmouths. In *Proceedings of the Australian Committee of the Association of Avian Veterinarians.* pp. 135–141. Dubbo.

[19] Rose, K. (2005). Common diseases of urban wildlife. Birds, Part 2. Report by The Australian Registry of Wildlife Health. Co-sponsored by the Western Plains Zoo, Dubbo and Taronga Zoo, Sydney. p. 1.

[20] OIE Working Group (2005). Report of the Meeting of the OIE Working Group on Wildlife Diseases. 73rd General Sessions, International Committee, HQ: World Organisation for Animal Health, Paris, 22–27 May 2005. Satellite meeting 14–16 February 2005 (web-posted), Paris; Gelis, S., Spratt, D.M. and Raidal, S.R. (2011). Neuroangiostrongyliasis and

other parasites in tawny frogmouths (*Podargus strigoides*) in south-eastern Queensland. *Australian Veterinary Journal* 89: 47–50; Blair, N.F., Orr, C.F., Delaney, A.P. and Herkes, G.K. (2013). Angiostrongylus meningoencephalitis: survival from minimally conscious state to rehabilitation. *Medical Journal of Australia* 198: 440–442.

21 Spratt, D.M. (2005). Neuroangiostrongyliasis: disease in wildlife and humans. *Microbiology Australia* 26 (June): 63–64.

22 Interestingly, Charles[18] noted that tawny frogmouths with OC toxicity abandon their nocturnal habits and are active during the day. Signs include weakness, inability to fly and convulsions. The birds can suffer from eye problems, show dilated pupils or keep their eyes closed, or even suffer from blindness.

23 Gelis, S., Spratt, D.M. and Raidal, S.R. (2011). Neuroangiostrongyliasis and other parasites in tawny frogmouths (*Podargus strigoides*) in south-eastern Queensland. *Australian Veterinary Journal* 89: 47–50; Aghazadeh, M., Jones, M.K., Aland, K.V., Reid, S.A., Traub, R.J., McCarthy, J.S. and Lee, R. (2015). Emergence of neural angiostrongyliasis in eastern Australia. *Vector-Borne and Zoonotic Diseases* 15 (3): 184–190.

24 Körtner, G. and Geiser, F. (1999). Roosting behaviour of the tawny frogmouth (*Podargus strigoides*). *Journal of Zoology, London* 248: 501–507.

25 Dickison, D.J. (1930). Unusual nesting sites. *Emu* 30: 145–147.

26 Baldwin, M. (1975). Birds of Inverell District, NSW. *Emu* 75: 113–120; Ford, H.A. and Bell, H. (1981). Density of birds in eucalypt woodland affected to varying degrees of dieback. *Emu* 81: 202–208; Keynes, T. (1987). Breeding the tawny frogmouth in captivity. *Bird Keeping in Australia* 30 (3): 33–37.

27 Doucette, L.I. (2010). Home range and territoriality of Australian owlet-nightjars *Aegotheles cristatus* in diverse habitats. *Journal of Ornithology* 151: 673–685.

28 Zdenek, C.N. (2017). A prolonged agonistic interaction between two Papuan frogmouths *Podargus papuensis*. *Australian Field Ornithology* 34: 26–29.

29 Kaplan, G. and Rogers, L.J. (2001). *Birds. Their Habits and Skills*. Allen and Unwin, Sydney.

30 Thomson, D.F.F. (1923). Notes on the tawny frogmouth (*Podargus strigoides*). *Emu* 22: 307–309.

31 Grevis, S. (Ed.) (1999). Out with the poison, in with the owls. *Natural Heritage, The Journal of the Natural Heritage Trust* 5: 9.

Chapter 6

1 Dunn, P.O. and Cockburn, A. (1999). Extrapair mate choice and honest signalling in cooperatively breeding superb fairy-wrens. *Evolution* 53: 938–946.

2 Doucette, L.I. (2010). Home range and territoriality of Australian owlet-nightjars *Aegotheles cristatus* in diverse habitats. *Journal of Ornithology* 151: 673–685.

3 Rieger, G. and Savin-Williams, R.C. (2012). The eyes have it: sex and sexual orientation differences in pupil dilation patterns. *PloS ONE* 7 (8): e40256.

4 Dell, J. (1971). Notes on the tawny frogmouth. *Western Australian Naturalist* 12: 21–22.

5 Cook, L.C. (1915). Notes from Poowong. *Emu* 15: 52–53; Florence, I. (1927). The curious frogmouth. *Emu* 27: 38–41; Tarr, H.E. (1985). Notes on nesting tawny frogmouths. *Australian Bird Watcher* 11 (2): 62–63; Higgins, P.J. (Ed.) (1999). Tawny frogmouths. In *Handbook of Australian, New Zealand and Antarctic Birds. Volume 4: Parrots to Dollarbirds*. pp. 964–983. Oxford University Press, Melbourne.

6   Schodde, R. and Mason, I.J. (1980). *Nocturnal Birds of Australia*. Lansdowne Editions, Melbourne.

7   Bright, J. (1935). Notes on a few birds of the Rochester District. *Emu* 34: 293–302.

8   Holyoak, D.T. (1999). Family Podargidae (Frogmouths). In *Handbook of the Birds of the World, Vol. 5, Barn-owls to Hummingbirds*. (Eds J. del Hoyo, A. Elliott and J. Sargantal) pp. 266–287. Lynx Edicions, Barcelona.

9   Tarr, H.E. (1985). Notes on nesting tawny frogmouths. *Australian Bird Watcher* 11 (2): 62–63.

10 Rae, S. and Rae, D. (2013). Orientation of tawny frogmouth (*Podargus strigoides*) nests and their position on branches optimises thermoregulation and cryptic concealment. *Australian Journal of Zoology* 61 (6): 469–474.

11 Schodde, R. and Mason, I.J. (1997). Aves (Columbidae to Coraciidae). In *Zoological Catalogue of Australia. Vol. 37.2*. (Eds W.W.K. Houston and A. Wells.) CSIRO Publishing, Melbourne; Higgins, P.J. (Ed.) (1999). Tawny frogmouths. In *Handbook of Australian, New Zealand and Antarctic Birds. Volume 4: Parrots to Dollarbirds*. pp. 964–983. Oxford University Press, Melbourne.

12 Brooker, M.G., Ridpath, M.G., Estbergs, A.J., Bywater, J., Hart, D.S. and Jones, M.S. (1979). Bird observations on the north-western Nullarbor Plain and neighbouring regions, 1967–1978. *Emu* 79: 176–190.

13 Florence, I. (1927) The curious frogmouth. *Emu* 27 (1): 38–41.

14 Marchant, S. (1983). Suggested nesting association between leaden flycatchers and noisy friarbirds. *Emu* 83: 119–122; Kaplan, G. (2019). *Australian Magpie*. 2nd edn. CSIRO Publishing, Melbourne.

15 Quinn, J.L. and Ueta, M. (2008). Protective nesting associations in birds. *Ibis* 150 (Suppl. 1): 146–167; Poka, M. (2014). Protective nesting association between the barred warbler *Sylvia nisoria* and the red-backed shrike *Lanius collurio*: an experiment using artificial and natural nests. *Ecological Research* 29 (5): 949–957.

16 Rae, S. (2013). Probable protective nesting association between Australasian figbird, noisy friarbird and Papuan frogmouth. *Australian Field Ornithology* 30: 126–130.

17 Fitzsimons, J. (2001). Tawny frogmouths displacing common mynas from a nesting hollow, and related observations. *Australian Bird Watcher* 19 (4): 129–131.

18 Higgins, P.J. (Ed.) (1999). Tawny frogmouths. In *Handbook of Australian, New Zealand and Antarctic Birds. Volume 4: Parrots to Dollarbirds*. pp. 964–983. Oxford University Press, Melbourne.

19 Bryant, C.E. (1942). Makeshift nest for frogmouths. *Emu* 42: 212; Ingram, G.J. (1994). Tawny frogmouth. In *Cuckoos, Nightbirds and Kingfishers of Australia*. (Ed. R. Strahan) pp. 88–90. Angus and Robertson, Sydney.

20 Körtner, G. and Geiser, F. (1999). Nesting behaviour and juvenile development of the tawny frogmouth *Podargus strigoides*. *Emu* 99: 212–217.

21 Thomas, B. (1957). Tawny frogmouth in captivity. *South Australian Ornithologist* 22: 46–47; Keynes, T. (1987). Breeding the tawny frogmouth in captivity. *Bird Keeping in Australia* 30 (3): 33–37.

22 Thomson, D.F.F. (1923). Notes on the tawny frogmouth (*Podargus strigoides*). *Emu* 22: 307–309; Fleay, D.H. (1925). The boobook owl and tawny frogmouth. *Emu* 25: 91–93; Hollands, D. (1991). *Birds of the Night: Owls, Frogmouths and Nightjars of Australia*. Reed Books, Balgowlah, New South Wales.

23 Gould, J. (1848). *The Birds of Australia 1840–1848. Vol. 2*. Self-published, London; Schodde, R. and Mason, I.J. (1980). *Nocturnal Birds of Australia*. Lansdowne Editions, Melbourne.

24 Cleere, N. (1998). *Nightjars: A Guide to Nightjars and Related Nightbirds of the World*. Pica Press, Robertsbridge, England.

25 Barnard, E.D. (1913). Birds and frogs. *Emu* 12: 193–194; Tarr, H.E. (1985). Notes on nesting tawny frogmouths. *Australian Bird Watcher* 11 (2): 62–63; Körtner, G. and Geiser, F. (1999). Nesting behaviour and juvenile development of the tawny frogmouth *Podargus strigoides*. *Emu* 99: 212–217.

26 Van Dyck, S. (2004). Tawdry frogmyths. *Nature Australia* (Winter): 20–21.

Chapter 7

1 Stoleson, S.H. (1999). The importance of early onset of incubation for the maintenance of egg viability. In *Proceedings of the 22nd International Ornithological Congress* 16–22 August 1998. (Eds N.J. Adams and R.H. Slotow) Birdlife, Durban, South Africa.

2 Viñuela, J. and Carrascal, L.M. (1999). Hatching patterns in precocial birds: a preliminary comparative analysis. In *Proceedings of the 22nd International Ornithological Congress* 16–22 August 1998. (Eds N.J. Adams and R.H. Slotow) Birdlife, Durban, South Africa.

3 Krebs, E.A., Cunningham, R.B. and Donnelly, C.F. (1999). Complex patterns of food allocation in asynchronously hatching broods of crimson rosellas. *Animal Behaviour* 57: 753–763.

4 Legge, S. (2000). Siblicide in the cooperative breeding laughing kookaburra (*Dacelo novaeguineae*). *Behavioural Ecology and Sociobiology* 48: 293–302.

5 Higgins, P.J. (Ed.) (1999). Tawny frogmouths. In *Handbook of Australian, New Zealand and Antarctic Birds. Volume 4: Parrots to Dollarbirds*. pp. 964–983. Oxford University Press, Melbourne; Schodde, R. and Mason, I.J. (1997). Aves (Columbidae to Coraciidae). In *Zoological Catalogue of Australia. Vol. 37.2*. (Eds W.W.K. Houston and A. Wells.) CSIRO Publishing, Melbourne.

6 Wilson, F.E. (1912). Oologists in the Mallee. *Emu* 12: 30–39.

7 O'Connor, R.J. (1984). *The Growth and Development of Birds*. John Wiley and Sons, Chichester; Rahn, H., Paganelli, C.V. and Ar, A. (1975). Relation of avian egg to body weight. *Auk* 92: 750–765; Ricklefs, R.E. and Starck, J.M. (1998). Embryonic growth and development. In *Avian Growth and Development*. (Eds J.M. Starck and R.E. Ricklefs) pp. 31–58. Oxford University Press, New York.

8 Williams, T.D. (1994). Intraspecific variation in egg size and egg composition in birds: effects on offspring fitness. *Biological Reviews* 68: 35–59.

9 Ricklefs, R.E. and Starck, J.M. (1998). Embryonic growth and development. In *Avian Growth and Development*. (Eds J.M. Starck and R.E. Ricklefs) pp. 31–58. Oxford University Press, New York.

10 Keynes, T. (1987). Breeding the tawny frogmouth in captivity. *Bird Keeping in Australia* 30 (3): 33–37; Fish, J.R. (1996). Notes on hand-rearing a tawny frogmouth at the Oklahoma City Zoological Park. *Avicultural Magazine* 102: 97–98; Körtner, G. and Geiser, F. (1999). Nesting behaviour and juvenile development of the tawny frogmouth *Podargus strigoides*. *Emu* 99: 212–217.

11 Fleay, D. (1968). *Nightwatchmen of Bush and Plain*. Jacaranda Press, Brisbane, Australia.

12 Schwabl, H. (1993). Yolk is a source of maternal testosterone for developing birds. *Proceedings of the National Academy of the Sciences USA* 90: 11444–11450; Schwabl, H. (1996). Maternal testosterone in the avian egg enhances postnatal growth. *Comparative Biochemistry and Physiology* 114A: 271–276.

13 Gil, D., Graves, J., Hazon, N. and Wells, A. (1999). Male attractiveness and differential testosterone investment in zebra finch eggs. *Science* 286: 126–128.

14 Körtner, G. and Geiser, F. (1999). Nesting behaviour and juvenile development of the tawny frogmouth *Podargus strigoides. Emu* 99: 212–217.

15 Oppenheim, R.W. (1974). The ontogeny of behavior in the chick embryo. Vol. 5. In *Advances in the Study of Behavior.* (Eds D.S. Lehram, J.S. Rosenblatt, R.A. Hinde and E. Shaw) pp. 133–171. Academic Press, New York.

16 Rogers, L.J. (1995). *The Development of Brain and Behaviour in the Chicken.* CAB International, Wallingford, Oxon, UK.

17 Tolhurst, B.E. and Vince, M.A. (1976). Sensitivity to odours in the embryo of the domestic fowl. *Animal Behaviour* 24: 772–779.

18 Lickliter, R. (1990). Premature visual stimulation accelerates intersensory functioning in bobwhite quail neonates. *Developmental Psychobiology* 23: 15–27.

19 Keynes, T. (1987). Breeding the tawny frogmouth in captivity. *Bird Keeping in Australia* 30 (3): 33–37.

20 Oppenheim, R.W. (1972). Prehatching and hatching behaviour in birds: a comparative study of altricial and precocial species. *Animal Behaviour* 20: 644–655.

21 Species of eucalypt with white flowers include *Eucalyptus tesselaris, E. confertiflora, E. apar-rerinja, E. calophylla* and *E. tetragonal.*

22 Hohtola, E. and Visser, G.H. (1998). Development of locomotion and endothermy in altricial and precocial birds. In *Avian Growth and Development.* (Eds J.M. Starck and R.E. Ricklefs) pp. 157–173. Oxford University Press, New York.

23 Holyoak, D.T. (2001). *Nightjars and their Allies.* Oxford University Press, New York.

24 Visser, G.H. (1998). Development of temperature regulation. In *Avian Growth and Development.* (Eds J.M. Starck and R.E. Ricklefs) pp. 117–156. Oxford University Press, New York.

25 Florence, I. (1927). The curious frogmouth. *Emu* 27: 39.

26 Apanius, V. (1998). Ontogeny of immune function. In *Avian Growth and Development.* (Eds J.M. Starck and R.E. Ricklefs) pp. 203–222. Oxford University Press, New York.

27 Lasiewski, R.C. and Bartholomew, G.A. (1966). Evaporative cooling in the poorwill and the tawny frogmouth. *Condor* 68: 253–262.

28 Tarr, H.E. (1985). Notes on nesting tawny frogmouths. *Australian Bird Watcher* 11 (2): 62–63.

29 Schodde, R. and Mason, I.J. (1980). *Nocturnal Birds of Australia.* Lansdowne Editions, Melbourne.

30 Kaplan, G. (2004). *Australian Magpie: Biology and Behaviour of an Unusual Songbird.* CSIRO Publishing, Melbourne.

31 Ashby, E. (1924). Podargus strigoides. *South Australian Ornithologist* 8 (1): 63–64.

32 Higgins, P.J. (Ed.) (1999). Tawny frogmouths. In *Handbook of Australian, New Zealand and Antarctic Birds. Volume 4: Parrots to Dollarbirds.* pp. 964–983. Oxford University Press, Melbourne.

33 Higgins, P.J. (Ed.) (1999). Tawny frogmouths. In *Handbook of Australian, New Zealand and Antarctic Birds. Volume 4: Parrots to Dollarbirds.* p. 969. Oxford University Press, Melbourne.

Chapter 8

1 Frauca, H. (1973). Tawny frogmouth. *Australian Birdlife* 1 (2): 19–22; Holyoak, D.T. (2001). *Nightjars and Their Allies.* Oxford University Press, New York.

2 Andrew, R.J. (1961). The displays given by passerines in courtship and reproductive fighting. A review. *Ibis* 103: 549–579.

3 Gregory, R. and Hopkins, P. (1974). Pupils of a talking parrot. *Nature* 252: 637–638.
4 Rogers, L.J. (1997). *Minds of Their Own: Thinking and Awareness in Animals*. Allen and Unwin, Sydney.
5 Kaplan, G. (2015). *Bird Minds. Cognition and Behaviour in Australian Native Birds*. CSIRO Publishing, Melbourne; see also Chevalier-Skolnikoff, S. (1973). Facial expression of emotion in nonhuman primates. In *Darwin and Facial Expression*. (Ed. P. Ekman) pp. 11–90. Academic Press, New York.
6 Bright, J. (1935). Notes on a few birds of the Rochester District. *Emu* 34: 293–302.
7 Rogers, L.J. and Bradshaw, J.L. (1996). Motor asymmetries in birds and nonprimate mammals. In *Manual Asymmetries in Motor Performance*. (Eds D. Elliott and E.A. Roy) pp. 3–31. CRC Press, New York; Hoese, W.J., Podos, J.E., Boetticher, N.C. and Nowicki, S.T. (2000). Vocal tract function in birdsong production: experimental manipulation of beak movements. *Journal of Experimental Biology* 203 (12): 1845–1855.
8 King, A.S. and McLelland, J. (1984). *Birds: Their Structure and Function*. 2nd edn. Balliere Tindall, Sussex, England; King, A.S. (1993). Apparatus respiratorius (systema respiratorium). In *Handbook of Avian Anatomy: Nomina Anatomica Avium*. (Eds J.J. Baumel *et al.*) Publications by the Nuttall Ornithological Club, No.23, Cambridge, Massachusetts. pp. 257–299.
9 Nottebohm, F. (1972). Neural lateralization of vocal control in a passerine bird. II. Subsong, calls and a theory of vocal learning. *Journal of Experimental Zoology* 179: 35–50.
10 Thorpe, W.H. (1972). Duetting and antiphonal song in birds. Its extent and significance. *Behaviour. An International Journal of Comparative Ethology*, Supplement XVIII. Leiden; Levin, R.N. (1996). Song behaviour and reproductive strategies in a duetting wren, *Thryothorus nigricapillus*. I. Removal experiments. *Animal Behaviour* 52: 1093–1106; Levin, R.N. (1996). Song behaviour and reproductive strategies in a duetting wren, *Thryothorus nigricapillus*. II. Playback experiments. *Animal Behaviour* 52: 1107–1117.
11 Higgins, P.J. (Ed.) (1999). Tawny frogmouths. In *Handbook of Australian, New Zealand and Antarctic Birds. Volume 4: Parrots to Dollarbirds*. pp. 964–983. Oxford University Press, Melbourne.
12 Davis, W.E. and Beehler, B.M. (1993). Dual singing between an adult and fledgling marbled frogmouth. *Corella* 17 (4): 111–113; Debus, S.J.S. (1997). Vocal behaviour of the southern boobook *Ninox novaeseelandiae* and other nocturnal birds. In *Australian Raptor Studies II: Proceedings, 2nd Australasia Raptor Association Conference*, Currumbin, Queensland, 8–9 April 1996. (Eds G. Czechura and S. Debus) Birds Australia Monograph 3, Hawthorn East, Melbourne.
13 Frauca, H. (1973). Tawny frogmouth. *Australian Birdlife* 1 (2): 19–22; Higgins, P.J. (Ed.) (1999). Tawny frogmouths. In *Handbook of Australian, New Zealand and Antarctic Birds. Volume 4: Parrots to Dollarbirds*. pp. 964–983. Oxford University Press, Melbourne.
14 Marshall, A.J. (1935). On the birds of the McPherson Ranges, Mt. Warning, and contiguous lowlands. *Emu* 35: 36–49.
15 Mattingley, A.H.E. (1910). Production of *Podargus* call. *Emu* 10: 246.
16 Slabbekoorn, H. and Cate, C.T. (1999). Collared dove responses to playback: slaves to the rhythm. *Ethology* 105: 377–391.
17 Hardouin, L.A., Bretagnolle, V., Tabel, P., Bavoux, C., Burneleau, G. and Reby, D. (2009). Acoustic cues to reproductive success in male owl hoots. *Animal Behaviour* 78 (4): 907–913; Appleby, B.M. and Redpath, S.M. (1997). Variation in the male territorial hoot of the tawny

owl *Strix aluco* in three English populations. *Ibis* 139 (1): 152–158; Riede, T., Eliason, C.M., Miller, E.H., Goller, F. and Clarke, J.A. (2016). Coos, booms, and hoots: the evolution of closed-mouth vocal behavior in birds. *Evolution* 70 (8): 1734–1746.

18 Stefanski, R.A. and Falls, J.B. (1972). A study of distress calls of song, swamp and white-throated sparrows (Aves: Fringillidae). I. Intraspecific responses and functions. *Canadian Journal of Zoology* 50: 1501–1512.

19 Marler, P. (1955). Characteristics of some animal calls. *Nature* 176: 6–8; Jurisevic, M.A. and Sanderson, K.J. (1994). Alarm vocalisations in Australian birds: convergent characteristics and phylogenetic differences. *Emu* 94: 69–77; Jurisevic, M.A. and Sanderson, K.J. (1998). A comparative analysis of distress call structure in Australian passerine and non-passerine species: influence of size and phylogeny. *Journal of Avian Biology* 29: 61–71.

20 Fleay, D. (1968). *Nightwatchmen of Bush and Plain*. Jacaranda Press, Brisbane, Australia.

21 Chisholm, A.H. (1934). *Bird Wonders of Australia*. Angus and Robertson, Sydney; Kaplan, G. (2000). Enchanting tawny frogmouths. *GEO* 22 (2): 45–50.

22 Frauca, H. (1973). Tawny frogmouth. *Australian Birdlife* 1 (2): 19–22.

Epilogue

1 Rose, K. (2005). Common Diseases of Urban Wildlife. Birds, Part 2. Report by The Australian Registry of Wildlife Health. Co-sponsored by the Western Plains Zoo, Dubbo and Taronga Zoo, Sydney. p. 1.

2 Loss, S.R., Ruiz, M.O. and Brawn, J.D. (2009). Relationships between avian diversity, neighborhood age, income, and environmental characteristics of an urban landscape. *Biological Conservation* 142 (11): 2578–2585.

References

Aghazadeh, M., Jones, M. K., Aland, K. V., Reid, S. A., Traub, R. J., McCarthy, J. S., and Lee, R. (2015). Emergence of neural angiostrongyliasis in eastern Australia. *Vector Borne and Zoonotic Diseases (Larchmont, N.Y.)* **15**(3), 184–190.

Andrew, R. J. (1961). The displays given by passerines in courtship and reproductive fighting. A review. *The Ibis* **103**, 549–579.

Apanius, V. (1998). Ontogeny of immune function. In *Avian Growth and Development*. (Eds J.M. Starck and R.E. Ricklefs) pp. 203–222. Oxford University Press, New York.

Appleby, B. M., and Redpath, S. M. (1997). Variation in the male territorial hoot of the tawny owl *Strix aluco* in three English populations. *The Ibis* **139**(1), 152–158.

Ashby, E. (1924). *Podargus strigoides. South Australian Ornithologist* **8**(1), 63–64.

Austin, O. L. (1963). Goatsucker-Caprimulgiformes. In *Birds of the World. A Survey of the 27 orders and 155 families*. pp. 159–163. Golden Press/Hamlyn Publishing House, Felham, Middlesex, England.

Baldwin, M. (1975). Birds of Inverell District, NSW. *Emu* **75**, 113–120.

Bang, B. G. (1971). Functional anatomy of the olfactory system in 23 orders of birds. *Acta Anatomica* **79**, 1–76.

Barker, R. D., and Vestjens, W. J. M. (1979). *The Food of Australian Birds 1. Non-Passerines*. CSIRO, Melbourne.

Barnard, E. D. (1913). Birds and frogs. *Emu* **12**, 193–194.

Barrett, G., Silcocks, A., Barry, S., Cunningham, R., and Poulter, R. (2003). *The New Atlas of Australian Birds*. Birds Australia, Melbourne.

Bech, C., and Nicol, S. C. (1999). Thermoregulation and ventilation in the tawny frogmouth, *Podargus strigoides*: a low-metabolic avian species. *Australian Journal of Zoology* **47**, 143–153.

Bennet, P. M., and Harvey, P. H. (1987). Active and resting metabolism in birds: allometry, phylogeny and ecology. *Journal of Zoology* **213**, 327–363.

Blair, N. F., Orr, C. F., Delaney, A. P., and Herkes, G. K. (2013). Angiostrongylus meningoencephalitis: survival from minimally conscious state to rehabilitation. *The Medical Journal of Australia* **198**, 440–442.

Bright, J. (1935). Notes on a few birds of the Rochester District. *Emu* **34**, 293–302.

Brooker, M. G., Ridpath, M. G., Estbergs, A. J., Bywater, J., Hart, D. S., and Jones, M. S. (1979). Bird observations on the north-western Nullarbor Plain and neighbouring regions, 1967–1978. *Emu* **79**, 176–190.

Bryant, C. E. (1942). Makeshift nest for frogmouths. *Emu* **42**, 212

Charles, J. A. (1995). Organochlorine toxicity in tawny frogmouths. In *Proceedings of the Australian Committee of the Association of Avian Veterinarians.* pp. 135–141. Dubbo.

Chevalier-Skolnikoff, S. (1973). Facial expression of emotion in nonhuman primates. In *Darwin and Facial Expression.* (Ed. P. Ekman) pp. 11–90. Academic Press, New York.

Chisholm, A. H. (1934). *Bird Wonders of Australia.* Angus and Robertson, Sydney.

Clark, L., and Mason, J. R. (1987). Olfactory discrimination of plant volatiles by the European starling. *Animal Behaviour* **35**, 227–235.

Cleere, N. (1998). *Nightjars: A Guide to Nightjars and Related Nightbirds of the World.* Pica Press, Robertsbridge, England.

Cleere, N., Kratter, A. W., Steadman, D. W., Braun, M. J., Huddleston, C. J., Filardi, C. E., and Dutson, G. (2007). A new genus of frogmouth (Podargidae) from the Solomon Islands – results from a taxonomic review of *Podargus ocellatus inexpectatus* Hartert 1901. *The Ibis* **149**, 271–286.

Cleijne, J. (2011). Tawny frogmouth study. *Wildlife Australia* **48**(2), 48.

Coleman, E. (1946). Foods of the tawny frogmouth. *Victorian Naturalist* **63**, 111–115.

Cook, L. C. (1915). Notes from Poowong. *Emu* **15**, 52–53.

Cooper, R. M., and McAllan, I. A. W. (1995). *The Birds of Western New South Wales: A Preliminary Atlas.* NSW Bird Atlassers Inc., Albury, NSW.

Cox, J. B., and Pedler, L. P. (1977). Birds recorded during three visits to the far north-east of South Australia. *South Australian Ornithologist* **27**, 231–250.

Davey, C., and Einoder, L. (2001). Predation of house mice by the tawny frogmouth *Podargus strigoides. Australian Bird Watcher* **19**(3), 103–104.

Davis, W. E., and Beehler, B. M. (1993). Dual singing between an adult and fledgling marbled frogmouth. *Corella* **17**(4), 111–113.

Debus, S. J. S. (1997). Vocal behaviour of the southern boobook *Ninox novaeseelandiae* and other nocturnal birds. In *Australian Raptor Studies II: Proceedings, 2nd Australasia Raptor Association Conference*, Currumbin, Queensland, 8–9 April 1996. (Eds G. Czechura and S. Debus) Birds Australia Monograph 3, Hawthorn East, Melbourne.

Dell, J. (1971). Notes on the tawny frogmouth. *Western Australian Naturalist (Perth)* **12**, 21–22.

Dickison, D. J. (1930). Unusual nesting sites. *Emu* **30**, 145–147.

Dickman, C. R. (2009). House cats as predators in the Australian environment: impacts and management. *Human–Wildlife Conflicts* **3**(1), 41–48.

Doucette, L. I. (2010). Home range and territoriality of Australian owlet-nightjars *Aegotheles cristatus* in diverse habitats. *Journal für Ornithologie* **151**, 673–685.

Doucette, L. I., Brigham, R. M., Pavey, C. R., and Geiser, F. (2012). Prey availability affects daily torpor by free-ranging Australian owlet-nightjars (*Aegotheles cristatus*). *Oecologia* **169**, 361–372.

Dowling, B., Seebeck, J. H., and Lowe, K. W. (1994). Technical Report. Arthur Rylah Institute of Environmental Research, Heidelberg, Victoria. p. 34.

Dunn, P. O., and Cockburn, A. (1999). Extrapair mate choice and honest signalling in cooperatively breeding superb fairy-wrens. *Evolution* **53**, 938–946.

Fish, J. R. (1996). Notes on hand-rearing a tawny frogmouth at the Oklahoma City Zoological Park. *Avicultural Magazine* **102**, 97–98.

Fitzsimons, J. (2001). Tawny frogmouths displacing common mynas from a nesting hollow, and related observations. *Australian Bird Watcher* **19**(4), 129–131.

Fleay, D. H. (1925). The boobook owl and tawny frogmouth. *Emu* **25**, 91–93.

Fleay, D. (1968). *Nightwatchmen of Bush and Plain*. Jacaranda Press, Brisbane.

Florence, I. (1927). The curious frogmouth. *Emu* **27**, 39.

Ford, J. R. (1986). Avian hybridization and allopatry in the region of the Einasleigh Uplands and Burdekin-Lynd Divide, north-eastern Queensland. *Emu* **86**, 87–110.

Ford, H. A., and Bell, H. (1981). Density of birds in eucalypt woodland affected to varying degrees of dieback. *Emu* **81**, 202–208.

Frauca, H. (1973). Tawny frogmouth. *Australian Birdlife* **1**(2), 19–22.

Gall, B. C., and Longmore, N. W. (1978). Avifauna of the Thredbo Valley, Kosciusko National Park. *Emu* **78**, 189–196.

Geiser, F. (2012). Hibernation. *Current Biology* **23**(5), 188–193.

Gelis, S., Spratt, D. M., and Raidal, S. R. (2011). Neuroangiostrongyliasis and other parasites in tawny frogmouths (*Podargus strigoides*) in south-eastern Queensland. *Australian Veterinary Journal* **89**, 47–50.

Gil, D., Graves, J., Hazon, N., and Wells, A. (1999). Male attractiveness and differential testosterone investment in zebra finch eggs. *Science* **286**, 126–128.

Gill, F., and Donsker, D. (Eds) (2017). *IOC World Bird List* (v. 7.2). doi :10.14344/IOC.ML.7.2

Gottschaldt, K.-M. (1985). Structure and function of avian somatosensory receptors. In *Form and Function in Birds, Vol. 3*. (Eds A.S. King and J. McLelland) pp. 375–461. Academic Press, London, New York.

Goujon, E. (1869). *Sur un appareil de corpuscules tactiles situe dans le bec des perroquets*. E. Martinet.

Gould, J. (1848). *The Birds of Australia 1840–1848. Vol. 2*. Self-published, London.

Gregory, R., and Hopkins, P. (1974). Pupils of a talking parrot. *Nature* **252**, 637–638.

Grevis, S. (Ed.) (1999). Out with the poison, in with the owls. *Natural Heritage. The Journal of the Natural Heritage Trust* **5**, 9.

Grubb, T. C. (1972). Smell and foraging in shearwaters and petrels. *Nature* **237**, 404–405.

Hackett, S. J., Kimball, R. T., Reddy, S., Bowie, R. C., Braun, E. L., Braun, M. J., Chojnowski, J. L., Cox, W. A., Han, K. L., Harshman, J., and Huddleston, C. J. (2008). A phylogenomic study of birds reveals their evolutionary history. *Science* **320**(5884), 1763–1768.

Hardouin, L. A., Bretagnolle, V., Tabel, P., Bavoux, C., Burneleau, G., and Reby, D. (2009). Acoustic cues to reproductive success in male owl hoots. *Animal Behaviour* **78**(4), 907–913.

Haselgrove, P. (1975). Notes on the birds of Groote Eylandt, N.T. *The Sunbird* **6**, 32–41.

Higgins, P. J. (Ed.) (1999). Tawny frogmouths. In *Handbook of Australian, New Zealand and Antarctic Birds. Volume 4: Parrots to Dollarbirds*. pp. 964–983. Oxford University Press, Melbourne.

Hindwood, K. A., and McGill, A. R. (1951). The 'Derra Derra' 1950 Camp-out of the R.A.O.U. *Emu* **50**, 217–218.

Hobbs, J. N. (1961). The birds of south-west New South Wales. *Emu* **61**, 21–55.

Hoese, W. J., Podos, J. E., Boetticher, N. C., and Nowicki, S. T. (2000). Vocal tract function in birdsong production: experimental manipulation of beak movements. *The Journal of Experimental Biology* **203**(12), 1845–1855.

Hogan F. E., Weaving M., Johnston G. R., and Gardner, M.G. (2012). Isolation and characterisation via 454 sequencing of microsatellites from the tawny frogmouth, *Podargus strigoides* (Class Aves, Family Podargidae). *Australian Journal of Zoology* **60**, 133–136.

Hohtola, E., and Visser, G. H. (1998). Development of locomotion and endothermy in altricial and precocial birds. In *Avian Growth and Development*. (Eds J.M. Starck and R.E. Ricklefs) pp. 157–173. Oxford University Press, New York.

Hollands, D. (1991). *Birds of the Night: Owls, Frogmouths and Nightjars of Australia*. Reed Books, Balgowlah, New South Wales.

Holyoak, D. T. (1999). Family Podargidae (Frogmouths). In *Handbook of the Birds of the World, Vol. 5, Barn-owls to Hummingbirds*. (Eds J. del Hoyo, A. Elliott and J. Sargantal) pp. 266–287. Lynx Edicions, Barcelona.

Holyoak, D. T. (2001). *Nightjars and Their Allies*. Oxford University Press, New York.

Hopkins, N. (1948). Birds of Townsville and district. *Emu* **47**, 331–347.

Ingram, G. J. (1994). Tawny frogmouth. In *Cuckoos, Nightbirds and Kingfishers of Australia*. (Ed. R. Strahan) pp. 88–90. Angus and Robertson, Sydney.

Iwaniuk, A. N., and Wylie, D. R. W. (2006). The evolution of stereopsis and the Wulst in caprimulgiform birds: a comparative analysis. *Journal of Comparative Physiology. A, Neuroethology, Sensory, Neural, and Behavioral Physiology* **192**, 1313–1326.

Jetz, W., Steffen, J., and Linsenmair, K. E. (2003). Effects of light and prey availability on nocturnal, lunar and seasonal activity of tropical nightjars. *Oikos* **103**, 627–639.

Jones, J. (1951). Victorian Hattah Lakes. *Emu* **52**, 225–254.

Jones, D. (1986). A survey of the bird community of 'Bullamon Plains' Thallon, Southern Queensland R.A.O.U. Camp-out. *The Sunbird* **16**(1), 1–11.

Jones, R. B., and Roper, T. J. (1997). Olfaction in the domestic fowl: a critical review. *Physiology & Behavior* **62**, 1009–1018.

Jurisevic, M. A., and Sanderson, K. J. (1994). Alarm vocalisations in Australian birds: convergent characteristics and phylogenetic differences. *Emu* **94**, 69–77.

Jurisevic, M. A., and Sanderson, K. J. (1998). A comparative analysis of distress call structure in Australian passerine and non-passerine species: influence of size and phylogeny. *Journal of Avian Biology* **29**, 61–71.

Kaplan, G. (2000). Enchanting tawny frogmouths. *GEO* **22**(2), 45–50.

Kaplan, G. (2004). *Australian Magpie: Biology and Behaviour of an Unusual Songbird.* CSIRO Publishing, Melbourne.

Kaplan, G. (2015). *Bird Minds: Cognition and Behaviour in Australian Native Birds.* CSIRO Publishing, Melbourne.

Kaplan, G. (2019). *Australian Magpie.* 2nd edn. CSIRO Publishing, Melbourne.

Kaplan, G., and Rogers, L. J. (2001). *Birds. Their Habits and Skills.* Allen and Unwin, Sydney.

Kaplan, G., and Rogers, L. J. (2006). Head-cocking as a form of exploration in marmosets and its development. *Journal of Developmental Psychobiology* **48**(2), 551–560.

Kaplan, G., and Rogers, L. J. (2013). Stability of referential signalling across time and locations: testing alarm calls of Australian magpies (*Gymnorhina tibicen*) in urban and rural Australia and in Fiji. *PeerJ* **1**, e112 10.7717/peerj.112.

Keynes, T. (1987). Breeding the tawny frogmouth in captivity. *Bird Keeping in Australia* **30**(3), 33–37.

King, A. S. (1993). Apparatus respiratorius (systema respiratorium). In *Handbook of Avian Anatomy: Nomina Anatomica Avium.* (Eds J.J. Baumel *et al.*) Publications by the Nuttall Ornithological Club, No.23, Cambridge, Massachusetts. pp. 257–299.

King, A. S., and McLelland, J. (1984). *Birds: Their Structure and Function.* 2nd edn. Balliere Tindall, Sussex, England.

Körtner, G., and Geiser, F. (1999). Roosting behaviour of the tawny frogmouth (*Podargus strigoides*). *Journal of Zoology* **248**, 501–507.

Körtner, G., and Geiser, F. (1999). Nesting behaviour and juvenile development of the tawny frogmouth *Podargus strigoides*. *Emu* **99**, 212–217.

Körtner, G., Brigham, R. M., and Geiser, F. (2001). Torpor in free-ranging frogmouths (*Podargus strigoides*). *Physiological and Biochemical Zoology* **74**(6), 789–797.

Krebs, E. A., Cunningham, R. B., and Donnelly, C. F. (1999). Complex patterns of food allocation in asynchronously hatching broods of crimson rosellas. *Animal Behaviour* **57**, 753–763.

Kühne, R., and Lewis, B. (1985). External and middle ears. In *Form and Function in Birds* (Eds A.S. King and J. McLelland) pp. 227–272. Academic Press, London.

Lasiewski, R. C., and Bartholomew, G. A. (1966). Evaporative cooling in the poorwill and the tawny frogmouth. *The Condor* **68**, 253–262.

Lasiewski, R. C., Dawson, W. R., and Bartholomew, G. A. (1970). Temperature regulation in the little Papuan frogmouth, *Podargus ocellatus*. *The Condor* **72**, 332–338.

Legge, S. (2000). Siblicide in the cooperative breeding laughing kookaburra (*Dacelo novaeguineae*). *Behavioral Ecology and Sociobiology* **48**, 293–302.

Levin, R. N. (1996). Song behaviour and reproductive strategies in a duetting wren, *Thryothorus nigricapillus*. II. Playback experiments. *Animal Behaviour* **52**, 1107–1117.

Lickliter, R. (1990). Premature visual stimulation accelerates intersensory functioning in bobwhite quail neonates. *Developmental Psychobiology* **23**, 15–27.

Lord, E. A. R. (1956). The birds of the Murphy's Creek District, southern Queensland. *Emu* **56**, 100–128.

Loss, S. R., Ruiz, M. O., and Brawn, J. D. (2009). Relationships between avian diversity, neighborhood age, income, and environmental characteristics of an urban landscape. *Biological Conservation* **142**(11), 2578–2585.

Loyn, R. H. (1985). Bird populations in successional forests of Mountain Ash *Eucalyptus regans* in Central Victoria. *Emu* **85**(4), 213–230.

Marchant, S. (1983). Suggested nesting association between leaden flycatchers and noisy friarbirds. *Emu* **83**, 119–122.

Marler, P. (1955). Characteristics of some animal calls. *Nature* **176**, 6–8.

Marshall, A. J. (1935). On the birds of the McPherson Ranges, Mt. Warning, and contiguous lowlands. *Emu* **35**, 36–49.

Martin, G. R. (1985). Eye. In *Form and Function in Birds. Vol. 3*. (Eds A.S. King and J. McLelland) pp. 311–373. Academic Press, London.

Martin, G. R. (1990). *Birds by Night*. T and A.D. Poyser, London.

Martin, G. R., and Katzir, G. (1999). Visual fields in short-toed eagles, *Ciraetus gallicus* (Accipitridae), and the function of binocularity in birds. *Brain, Behavior and Evolution* **53**, 55–66.

Mason, J., and Clark, L. (1995). Capsaicin detection in trained European starlings: the importance of olfaction and trigeminal chemoreception. *The Wilson Bulletin* **107**, 165.

Mattingley, A. H. E. (1910). Production of *Podargus* call. *Emu* **10**, 246.

McColl, W. S. (1929). Avifauna of the Hampton Tableland, Hampton Lowlands and Nullabor Plain. *Emu* **29**, 91–100.

McCulloch, E. M. (1975). Variations in the mass of captive tawny frogmouths. *The Australian Bird Bander* **13**(1), 9–11.

McFadden, S. A. (1993). Constructing the three-dimensional image. In *Vision, Brain, and Behavior in Birds*. (Eds H.P. Zeigler and H.-J. Bischof) pp. 47–61. The MIT Press, Cambridge, Massachusetts.

McKechnie, A. E., and Mzilikazi, N. (2011). Heterothermy in Afrotropical mammals and birds: a review. *Integrative and Comparative Biology* **51**(3), 349–363.

McNab, B. K., and Bonaccorso, F. J. (1995). The energetics of Australian swifts, frogmouths, and nightjars. *Physiological Zoology* **68**, 245–261.

Menzel, C. R., and Menzel, E. W. J. (1980). Head-cocking and visual exploration in marmosets (*Saguinus fuscicollis*). *Behaviour* **75**, 219–233.

Miles, F. A. (1998). The neural processing of 3-D visual information: evidence from eye movements. *The European Journal of Neuroscience* **10**, 811–822.

Morris, D. (1980). *Manwatching. A Field Guide to Human Behaviour.* Triad Granada, imprint of Jonathan Cape, London.

Morton N. J., Britton P., Palasanthiran P., Bye A., Sugo E., Kesson A., Ardern-Holmes S., and Snelling T. L. (2013). Severe hemorrhagic meningoencephalitis due to *Angiostrongylus cantonensis* among young children in Sydney, Australia. *Clinical Infectious Diseases* **57**, 1158–1161.

Nef, S., Allaman, I., Fiumelli, H., De Castro, E., and Nef, P. (1996). Olfaction in birds: differential embryonic expression of nine putative odorant receptor genes in the avian olfactory system. *Mechanisms of Development* **55**, 65–77.

Nottebohm, F. (1972). Neural lateralization of vocal control in a passerine bird. II. Subsong, calls and a theory of vocal learning. *The Journal of Experimental Zoology* **179**, 35–50.

O'Connor, R. J. (1984). *The Growth and Development of Birds.* John Wiley and Sons, Chichester.

OIE Working Group (2005). Report of the Meeting of the OIE Working Group on Wildlife Diseases. 73rd General Sessions, International Committee, HQ: World Organisation for Animal Health, Paris, 22–27 May 2005. Satellite meeting 14–16 February 2005 (web-posted), Paris.

Oppenheim, R. W. (1972). Prehatching and hatching behaviour in birds: a comparative study of altricial and precocial species. *Animal Behaviour* **20**, 644–655.

Oppenheim, R. W. (1974). The ontogeny of behavior in the chick embryo. Vol. 5. In *Advances in the Study of Behavior.* (Eds D.S. Lehram, J.S. Rosenblatt, R.A. Hinde and E. Shaw) pp. 133–171. Academic Press, New York.

Payne, R. S. (1962). How the barn owl locates prey by hearing. *Living Bird* **1**, 151–159.

Payne, R.S. (1971). Acoustic location of prey by barn owls (*Tyto alba*). *Journal of Experimental Biology* 54: 535–573; Martin, G.R. (1990). *Birds by Night.* T. and A.D. Poyser, London.

Payne, R. S. (1971). Acoustic location of prey by barn owls (*Tyto alba*). *The Journal of Experimental Biology* **54**, 535–573.

Pettigrew, J. D., and Konishi, M. (1976). Neurones selective for orientation and binocular disparity in the visual Wulst of the barn owl (*Tyto alba*). *Science* **193**, 675–678.

Poka, M. (2014). Protective nesting association between the barred warbler *Sylvia nisoria* and the red-backed shrike *Lanius collurio*: an experiment using artificial and natural nests. *Ecological Research* **29**(5), 949–957.

Quinn, J. L., and Ueta, M. (2008). Protective nesting associations in birds. *The Ibis* **150**(Suppl. 1), 146–167.

Rae, S. (2013). Probable protective nesting association between Australasian figbird, noisy friarbird and Papuan frogmouth. *Australian Field Ornithology* **30**, 126–130.

Rae, S., and Rae, D. (2013). Orientation of tawny frogmouth (*Podargus strigoides*) nests and their position on branches optimises thermoregulation and cryptic concealment. *Australian Journal of Zoology* **61**(6), 469–474.

Rahn, H., Paganelli, C. V., and Ar, A. (1975). Relation of avian egg to body weight. *The Auk* **92**, 750–765.

Rattenborg, N. C., Lima, S. L., and Amlaner, C. J. (1999). Facultative control of avian unihemispheric sleep under risk of predation. *Behavioural Brain Research* **105**, 163–172.

Rattenborg, N. C., Amlaner, C. J., and Lima, S. L. (2000). Behavioral, neurophysiological and evolutionary perspectives on unihemispheric sleep. *Neuroscience and Biobehavioral Reviews* **24**, 817–842.

Reinertsen, R. E. (1983). Nocturnal hypothermia and its energetic significance for small birds living in the arctic and subarctic regions. A review. *Polar Research* **1**, 269–284.

Ricklefs, R. E., and Starck, J. M. (1998). Embryonic growth and development. In *Avian Growth and Development*. (Eds J.M. Starck and R.E. Ricklefs) pp. 31–58. Oxford University Press, New York.

Ridley, E. (1985). A tale of a tawny frogmouth. *Bird Keeping in Australia* **28**, 144–145.

Riede, T., Eliason, C. M., Miller, E. H., Goller, F., and Clarke, J. A. (2016). Coos, booms, and hoots: the evolution of closed-mouth vocal behavior in birds. *Evolution* **70**(8), 1734–1746.

Rieger, G., and Savin-Williams, R. C. (2012). The eyes have it: sex and sexual orientation differences in pupil dilation patterns. *PLoS One* **7**(8), e40256.

Rogers, L. J. (1995). *The Development of Brain and Behaviour in the Chicken*. CAB International, Wallingford, Oxon, UK.

Rogers, L. J. (1997). *Minds of Their Own: Thinking and Awareness in Animals*. Allen and Unwin, Sydney.

Rogers, L. J., and Bradshaw, J. L. (1996). Motor asymmetries in birds and nonprimate mammals. In *Manual Asymmetries in Motor Performance*. (Eds D. Elliott and E.A. Roy) pp. 3–31. CRC Press, New York.

Rogers, L. J., and Kaplan, G. (1998). *Not Only Roars and Rituals: Communication in Animals*. Allen and Unwin, Sydney, Australia.

Rogers, L. J., Stafford, D., and Ward, J. P. (1993). Head cocking in galagos. *Animal Behaviour* **45**, 943–952.

Rojas, L. M., McNeil, R., Cabana, T., and Lachaoelle, P. (1997). Diurnal and nocturnal visual function in two tactile foraging waterbirds: the American White Ibis and the Black Skimmer. *The Condor* **99**, 191–200.

Rose, K. (2005). Common Diseases of Urban Wildlife. Birds, Part 2. Report by The Australian Registry of Wildlife Health. Co-sponsored by the Western Plains Zoo, Dubbo and Taronga Zoo, Sydney. p. 1.

Rose, A. B., and Eldridge, R. H. (1997). Diet of the tawny frogmouth *Podargus strigoides* in eastern New South Wales. *Australian Bird Watcher* **17**, 25–33.

Saunders, D. A., and Ingram, J. (1995). *Birds of Southwestern Australia*. Surrey Beatty and Sons, Chipping Norton, Australia.

Schaeffel, F., Howland, H. C., and Farkas, L. (1986). Natural accommodation in the growing chicken. *Vision Research* **26**, 1977–1993.

Schodde, R., and Mason, I. J. (1980). *Nocturnal Birds of Australia*. Lansdowne Editions, Melbourne.

Schodde, R., and Mason, I. J. (1997). Aves (Columbidae to Coraciidae). In *Zoological Catalogue of Australia. Vol. 37.2*. (Eds W.W.K. Houston and A. Wells.) CSIRO Publishing, Melbourne.

Schwabl, H. (1993). Yolk is a source of maternal testosterone for developing birds. *Proceedings of the National Academy of Sciences of the United States of America* **90**, 11444–11450.

Schwabl, H. (1996). Maternal testosterone in the avian egg enhances postnatal growth. *Comparative Biochemistry and Physiology* **114A**, 271–276.

Sedgwick, E. H. (1949). Observations on the lower Murchison R.A.O.U. Camp. September 1948. *Emu* **48**, 212–242.

Serventy, D. L. (1936). Feeding methods of Podargus. With remarks on the possible causes of its aberrant habits. *Emu* **36**, 74–90.

Slabbekoorn, H., and Cate, C. T. (1999). Collared dove responses to playback: slaves to the rhythm. *Ethology* **105**, 377–391.

Smith, P. (1984). The forest avifauna near Bega, New South Wales. 1. Differences between forest types. *Emu* **84**, 200–210.

Spratt, D. M. (2005). Neuroangiostrongyliasis: disease in wildlife and humans. *Microbiology Australia* **26**(June), 63–64.

Stefanski, R. A., and Falls, J. B. (1972). A study of distress calls of song, swamp and white-throated sparrows (Aves: Fringillidae). I. Intraspecific responses and functions. *Canadian Journal of Zoology* **50**, 1501–1512.

Stoleson, S. H. (1999). The importance of early onset of incubation for the maintenance of egg viability. In *Proceedings of the 22nd International Ornithological Congress 16–22 August 1998*. (Eds N.J. Adams and R.H. Slotow) Birdlife, Durban, South Africa.

Swanson, M., and Sanderson, K. J. (1999). Preliminary observations on acoustic perceptions of alarm calls and natural sounds by Australian owls and frogmouths. *South Australian Ornithologist* **33**, 51–55.

Tarr, H. E. (1985). Notes on nesting tawny frogmouths. *Australian Bird Watcher* **11**(2), 62–63.

Thomas, B. (1957). Tawny frogmouth in captivity. *South Australian Ornithologist* **22**, 46–47.

Thomson, D. F. F. (1923). Notes on the tawny frogmouth (*Podargus strigoides*). *Emu* **22**, 307–309.

Thorpe, W.H. (1972). Duetting and antiphonal song in birds. Its extent and significance. *Behaviour. An International Journal of Comparative Ethology*, Supplement XVIII. Leiden.

Tolhurst, B. E., and Vince, M. A. (1976). Sensitivity to odours in the embryo of the domestic fowl. *Animal Behaviour* **24**, 772–779.

Turner, J. R. (1992). Effect of wildfire on birds at Weddin Mountain, New South Wales. *Corella* **16**(3), 65–74.

Van Dyck, S. (2004). Tawdry frogmyths. *Nature Australia* (Winter): 20–21.

van Oosterzee, P. (1997). *Where Worlds Collide: The Wallace Line*. Reed Books, Victoria.

Vanderwilligen, R. F., Frost, B. J., and Wagner, H. (1998). Stereoscopic depth perception in the owl. *Neuroreport* **9**, 1233–1237.

Vestjens, W. J. M. (1973). Wildlife mortality on a road in New South Wales. *Emu* **73**, 107–112.

Viñuela, J., and Carrascal, L. M. (1999). Hatching patterns in precocial birds: a preliminary comparative analysis. In *Proceedings of the 22nd International Ornithological Congress 16–22 August 1998*. (Eds N.J. Adams and R.H. Slotow) Birdlife, Durban, South Africa.

Visser, G. H. (1998). Development of temperature regulation. In *Avian Growth and Development*. (Eds J.M. Starck and R.E. Ricklefs) pp. 117–156. Oxford University Press, New York.

Wallman, J., and Pettigrew, J. D. (1985). Conjugate and disjunctive saccades in two avian species with contrasting oculomotor strategies. *The Journal of Neuroscience* **5**(6), 1418–1428.

Weaving, M. J., White, J. G., Isaac, B., and Cooke, R. (2011). The distribution of three nocturnal bird species across a suburban–forest gradient. *Emu* **111**, 52–58.

Weaving, M. J., White, J. G., Hower, K., Isaac, B., and Cooke, R. (2014). Sex-biased space-use response to urbanization in an endemic urban adapter. *Landscape and Urban Planning* **130**, 73–80.

Weaving, M. J., White, J. G., Isaac, B., Rendall, A. R., and Cooke, R. (2016). Adaptation to urban environments promotes high reproductive success in the tawny frogmouth (*Podargus strigoides*), an endemic nocturnal bird species. *Landscape and Urban Planning* **150**, 87–95.

Wild, J. M., Kubke, M. F., and Peña, J. L. (2008). A pathway for predation in the brain of the barn owl (*Tyto alba*). *The Journal of Comparative Neurology* **509**(2), 156–166.

Williams, T. D. (1994). Intraspecific variation in egg size and egg composition in birds: effects on offspring fitness. *Biological Reviews of the Cambridge Philosophical Society* **68**, 35–59.

Wilson, F. E. (1912). Oologists in the Mallee. *Emu* **12**, 30–39.

Woinarski, J. C. Z., Press, A. J., and Russell-Smith, J. (1989). The bird community of a sandstone plateau monsoon forest at Kakadu National Park, Northern Territory. *Emu* **89**, 223–231.

Zdenek, C. N. (2017). A prolonged agonistic interaction between two Papuan frogmouths *Podargus papuensis*. *Australian Field Ornithology* **34**, 26–29.

Index

www.ingramcontent.com/pod-product-compliance
Lightning Source LLC
Chambersburg PA
CBHW041118280326
41928CB00060B/3456